10×1/11

CHARLES DARWIN

Blackwell Great Minds

Edited by Steven Nadler

The Blackwell Great Minds series gives readers a strong sense of the fundamental views of the great western thinkers and captures the relevance of these figures to the way we think and live today.

1   *Kant* by Allen W. Wood
2   *Augustine* by Gareth B. Matthews
3   *Descartes* by André Gombay
4   *Sartre* by Katherine J. Morris
5   *Charles Darwin* by Michael Ruse
6   *Schopenhauer* by Robert Wicks

Forthcoming

Aristotle by Jennifer Whiting
Nietzsche by Richard Schacht
Plato by Paul Woodruff
Spinoza by Don Garrett
Wittgenstein by Hans Sluga
Heidegger by Taylor Carman
Maimonides by Tamar Rudavsky
Berkeley by Margaret Atherton
Leibniz by Christa Mercer
Shakespeare by David Bevington
Hume by Stephen Buckle
Kierkegaard by M. Jamie Ferreira
Mill by Wendy Donner and Richard Fumerton
Camus by David Sherman
Socrates by George H. Rudebusch
Hobbes by Edwin Curley
Locke by Gideon Yaffe

CHARLES DARWIN

Michael Ruse

Blackwell
Publishing

BLACKWELL PUBLISHING
350 Main Street, Malden, MA 02148-5020, USA
9600 Garsington Road, Oxford OX4 2DQ, UK
550 Swanston Street, Carlton, Victoria 3053, Australia

First published 2008 by Blackwell Publishing Ltd

1   2008

Library of Congress Cataloging-in-Publication Data

Ruse, Michael.
    Charles Darwin / Michael Ruse.
        p. cm. — (Blackwell great minds ; 4)
    Includes bibliographical references and index.
    ISBN 978-1-4051-4912-9 (hardcover : alk. paper)
    ISBN 978-1-4051-4913-6 (pbk. : alk. paper)   1. Evolution.   I. Title.
    B818.R86 2007
    576.8092—dc22

                                                                2007014520

A catalogue record for this title is available from the British Library.

Set in 10.5/13pt Galliard
by Graphicraft Limited, Hong Kong
Printed and bound in Singapore
by Utopia Press Pte Ltd

The publisher's policy is to use permanent paper from mills that operate a sustainable forestry policy, and which has been manufactured from pulp processed using acid-free and elementary chlorine-free practices. Furthermore, the publisher ensures that the text paper and cover board used have met acceptable environmental accreditation standards.

For further information on
Blackwell Publishing, visit our website:
www.blackwellpublishing.com

To Edward O. Wilson

Contents

List of Figures   viii
Preface   x

 1   Charles Darwin   1
 2   *On the Origin of Species*   21
 3   One Long Argument   54
 4   Neo-Darwinism   75
 5   The Consilience: One   99
 6   The Consilience: Two   122
 7   Humans   158
 8   Knowledge   191
 9   Morality   215
10   Religious Belief   241
11   The Origins of Religion   265
12   The Darwinian Revolution   287

Bibliography   308
Index   324

Figures

1.1	Darwin's illustration of his coral reef theory	6
1.2	Distribution of Galapagos tortoises	9
2.1	Varieties of fancy pigeon, originating in the rock pigeon of India	22
2.2	Sexual selection: *Lophornis ornatus* female and male	28
2.3	The "Tree of Life" as illustrated by Ernst Haeckel	32
2.4	The fossil record as known at the time of Darwin's *On the Origin of Species*	44
2.5	Vertebrate homologies	50
4.1	Mendel's first law	84
4.2	Distributions of the sickle-cell gene and types of malaria in Uganda (*c*.1949)	91
4.3	Adaptive landscape	98
5.1	Relationships in Hymenoptera	106
5.2	Matrix of hawks versus doves in an evolutionarily stable strategy	107
5.3	The age of earth	113
5.4	Skeletons of Archaeopteryx and pigeon	115
5.5	Titanothere	117
5.6	The Phanerozoic history of the taxonomic diversity of marine animal families	118
6.1	Ring of races in the greenish warbler complex	124
6.2	Wallace's Line	127
6.3	Plate tectonics	128
6.4	Biogeographical distribution	129
6.5	The great American interchange	131
6.6	Linnaean hierarchy	133

6.7	Cladograms used to establish classifications	139
6.8	The history of life	142
6.9	Stegosaurus	145
6.10	Spandrel from the Villa Farnesina, Rome	147
6.11	How the elephant got its trunk	148
6.12	Skeleton of the Irish elk	150
6.13	Pine cone: an example of an 8, 13 phyllotaxis	153
6.14	Molecular homologies	156
7.1	Darwin as an ape	161
7.2	Human evolution and relation to the higher apes	167
7.3	Saber-toothed mammals	184
7.4	Comparison of the skeletons of the tiger and the whale	188

Preface

Why include Charles Darwin, the father of evolutionary theory, in a series devoted to the thinking of great philosophers? There are several reasons, starting with the fact that Darwin was always interested in philosophy. As a young man he mixed with, and talked to, people working as philosophers, notably the historian and philosopher of science William Whewell; he read and thought hard about works by several of the masters, including Plato, Aristotle, Hume, and Kant, as well as many books by lesser thinkers; and he wrote on philosophical issues when they came within his scientific purview. For this reason alone it is no great surprise that much that Darwin had to say has considerable importance to the person interested in philosophy.

There are, however, stronger reasons for including Darwin in a series on philosophers. His work in itself is something that calls for philosophical analysis and that bears on philosophical issues. Thanks to Darwin, we now know that organisms were not created in six days miraculously, but are the end result of a long, slow, undirected process of natural change – evolution. Such a theory needs examining conceptually, to see how it is structured and how it makes its claims. And since the theory extends to humankind – we were not the result of a burst of creative activity at the end of the divine workweek – Darwin's thinking must also be explored for its implications for important questions in philosophy, both the theory of knowledge (epistemology) and the theory of morality (ethics).

These factors are the reasons for, and the themes of, this book. This is a book about Charles Darwin written for those who want to know about him and his work and their connections with, and implications for, philosophy. These intentions lie behind decisions made in the book about what topics to discuss and in what order. Although Darwin wrote many books, without apology I focus on his *Origin of Species* (with some later discussion of *The Descent of Man*), because that is where the philosophical issues arise. Because Darwin was, after all, first and foremost a great scientist, I take my first task to be that of saying what he said as a scientist, and my second task to be that of showing how the science has developed from his day to ours. If it turns out that Darwin's theory is simply wrong or inadequate as judged by the scientific standards of today, it would not necessarily mean that there is nothing of lasting philosophical importance in his thinking – we do not turn from Kant because he was mistaken in his belief that Newtonian mechanics is necessarily true – but it will affect the ways in which we judge its level of philosophical importance. Demonstrating that, however we judge the science, nothing need go to waste, as I give my exposition I will also show how the science itself gives rise to problems of considerable philosophical interest.

Then I will go on to look at the implications of Darwin's thought for traditional philosophical questions about knowledge, about morality, and as appropriate (that is, as it impinges on philosophy) about religion. Just as a book on Aristotle might legitimately move on to consider his influence on later philosophical thinkers like Aquinas, so (especially given that Darwin, unlike Aristotle, was not centrally a philosopher) I think it legitimate (and, indeed, highly desirable) to go on to consider the influence of Darwin on later philosophical thinkers, down to this day. Although I have striven to give a full and fair account of different opinions, I do not hesitate to give my views on what I think are the right positions to take.

I should say that this book is a labor of love. I have been fascinated now by Darwin for over four decades. To be able to put together my ideas and to draw my conclusions is both exciting and a privilege. The book is also one with a mission. When I began life as a professional philosopher, around the time I became interested in Darwin, one had to search long and far for any philosopher who thought that Darwin and his work had any significance for our subject. Most agreed with a comment made in the *Philosophical Investigations* by Ludwig Wittgenstein: "Darwin's theory has no more relevance for philosophy than any other hypothesis in natural science." I thought that was wrong then, and I think it is wrong now. Since the 1960s, things have changed a great deal. Many philosophers of science have turned to things Darwinian to analyze them and many others have brought Darwin into their philosophical discussions. But there is still a long way to go, and much work must done to overcome the hostility many philosophers still feel for the whole Darwinian enterprise. As the wife of the Bishop of Worcester is supposed to have said: "Descended from monkeys? My dear, let us hope that it is not true. But if it is true, let us hope that it not become widely known." It should be widely known and it should be the starting point of much in philosophy.

I want to thank Joe Cain, Peter Loptson, and Richard Richards, who read a draft version of this work and who gave me really good feedback. More generally, in my life as a scholar, I have been fortunate to have had really good friends and really ferocious enemies. I am never quite sure from which group I have learnt more. In a way, this book is dedicated to everybody. But there is one person, a really good friend, who stands out as the person who has truly enriched my thinking about evolution in more ways than I suspect I am aware. It is a small return to be able to put at the head of this book the name of Edward O. Wilson.

Michael Ruse
Tallahassee, Florida
December 2007

I am grateful to the following for permission to reproduce illustrations:

Figure 4.2: Allison, A. C. "Protection by the sickle-cell trait against subtertian malarial infection." *British Medical Journal* 1 (1954), p. 290, figures 1 and 2.

Figure 5.6: from J. J. Sepkoski. "A kinetic model of Phanerozoic taxonomic diversity, III." *Paleobiology* 10 (1984), p. 249, figure 1.

Figure 6.1: D. E. Irwin, S. Bensch, and T. D. Price. "Speciation in a Ring." *Nature* 409 (2001), pp. 333–7, figure 1 (p. 334).

Figure 6.14: S. B. Carroll, J. K. Grenier, and S. D. Weatherbee. *From DNA to Diversity: Molecular Genetics and the Evolution of Animal Design* (2001). Oxford: Blackwell Science, figure 2.8.

1

Charles Darwin

The English £10 note (about $20) carries a picture of Queen
Elizabeth on the front and, on the back, the picture of an old man,
with a wonderfully full beard. Every English child knows whose por-
trait this is, even if many are not really quite sure why he is so famous.
It is the picture of Charles Darwin, one of the truly great scientists
of all time. Let us learn something about him.

The early years

Charles Robert Darwin was born on February 12, 1809, in the
English midlands town of Shrewsbury (the first syllable pronounced
to rhyme with "blows" not "blues"), the same day as Abraham
Lincoln across the Atlantic (Browne 1995, 2002). He died at home,
in the Kentish village of Downe, on April 19, 1882. He was the fourth
of five children, the second of two sons, of Dr. Robert Darwin.
His paternal grandfather was Dr. Erasmus Darwin, a physician,
who died before his birth. Erasmus Darwin was an eighteenth-
century figure, well known not just for his skill at medicine (poor,
mad King George III tried to get him to come to court), but also
for his interest in science and technology (King-Hele 1963). He was
one of a number of inventors and businessmen – including Matthew
Boulton (the industrialist) and his partner James Watt (inventor and
improver of the steam engine), Joseph Priestley (the chemist), Samuel
Galton (the gun maker), and William Withering (botanist and discoverer
of digitalis) – who were members of the so-called Lunar Society, which
met once a month to discuss matters of science and technology and

their application to industrial questions. Erasmus Darwin was also a poet and an evolutionist. He believed that all organisms come from (probably) one original form, and then develop through time into the different kinds that are revealed from the past and around us today. Poetry and evolution often overlapped in the world of Erasmus Darwin, for he was much given to expressing his scientific speculations in verse.

Robert Darwin was a physician like his father, at least as well known and respected for his knowledge and his skills. Dr. Darwin was also a very important money man. Given his wide clientele, he was in a perfect position to bring together aristocrats in need of cash and with lands to mortgage and industrialists with cash to loan and seeking safe investments. As is common in these cases, then and now, this proved very profitable for the middleman, who was soon in the moneylending business himself. Even more wealth poured into the Darwin family from Charles's maternal grandfather, Josiah Wedgwood (a friend of Erasmus Darwin and a fellow "lunatik"), the man who brought the Industrial Revolution to the pottery business, learning and applying the Asian techniques in making what became known as bone china. The marriage settlement of Darwin's mother was very significant.

It is worth emphasizing these points, because at once we can start to put young Charles into context. He was not an aristocrat, but he was a member of the rich, upper-middle classes, the people who had done (and continued to do) very well out of the Industrial Revolution. One would expect him to be a solid citizen with a strong vested interest in his country and to appreciate its overall stability; yet probably more of a liberal, favorable to the innovations that machines and factories were bringing to Britain, than a conservative, who deplored every change to life as it had been in the eighteenth century and earlier; a man who favored reform but not revolt. One would also expect him to be happy with his lot, and not about to reject or repudiate it. In other words, however much of a scientific revolutionary Darwin was to be – and I believe he was a very great scientific revolutionary – he would not be like the Christian God, creating things from nothing. One would expect – and the expectation is fully realized – that Darwin would take what was given and (rather like a kaleidoscope) make of it a new picture. To understand Darwin is – as any evolutionist would have forecast – to understand his past and his influences.

Charles Darwin's genius was always more of the creative-bright than the IQ-bright variety. He was an indifferent student at school, where merit went to those clever at writing Greek and Latin verses or mastering the intricacies of Euclidean geometry. From an early age, however, he was interested in science. Charles and his older brother, also called Erasmus, used to do simple home experiments in chemistry. Given the great importance of that particular science in technological applications of pure theory, this was just what one might have expected of two children of the Industrial Revolution. Expecting to follow in the family profession, at a young age (16) Charles was packed off to Edinburgh, then the home of the finest medical school in Europe. Two years later – revolted by the operations and bored by the professors – he had had enough. Following his own inclinations, he had taken up natural history with a vengeance, but this did not compensate for living with the Scots in their gloomy capital. His family therefore redirected him to the perfect career for a young Englishman with considerable wealth and little obvious talent. He was to become an Anglican (that is, an Episcopalian) clergyman. To achieve this, Darwin had to have a degree from an English university. And so, early in 1828, Charles Darwin enrolled at Christ's College, at the University of Cambridge.

As someone who was already starting to show an interest in science, this was a good time for Darwin to go to Cambridge. Although there was no formal science teaching, a number of the professors were becoming very interested in science and were willing to admit to their number young men who shared their enthusiasms. Darwin soon became friendly with John Henslow, the professor of botany, Adam Sedgwick, the professor of geology, and William Whewell, then the professor of mineralogy but later to become the professor of philosophy (a career change one doubts has been replicated that frequently). Although these professors had no specific obligations in pursuit their subjects – and hitherto no incumbents had felt obliged to pursue them actively – now people were beginning to explore the world of nature and to marvel at its wonders. One should add that, at Cambridge, this was always done in a religious context and generally involved looking at nature to praise the abilities of the Creator. In those days, a professor at an English university (Oxford being the only other) had to be an ordained member of the Church of England (Anglican).

Darwin fit happily into this group, not only because of the science, but because at that early date he himself had no qualms whatsoever about the truths of Christianity or the Thirty-Nine Articles, subscription to which was a necessary condition for those who belonged to the state church. He clearly impressed his seniors because, when he graduated in 1831, through the connections of the Cambridge science group, he was given the opportunity to spend several years on board HMS *Beagle*, a British warship that was to chart the coasts around South America. Postponing his clerical career – the prospect of which was never formally repudiated but gradually and gently faded away – Darwin ended by spending five years on the ship, eventually circumnavigating the globe, before it returned to England in the autumn of 1836. Originally his status on the ship was primarily that of a companion to the captain, but rapidly he became the ship's naturalist, and spent a considerable time studying the flora and fauna of the lands that he visited, sending massive collections back home for study by the appropriate specialists.

Charles Darwin was to make his great mark as a biologist, but in the early years he focused more on geological questions (Herbert 2005). Around 1830, as Darwin started to enter the ranks of professional scientist, geology was a science with a significant profile, if only because of its commercial importance. Road building, canal digging, mining – all of these were essential activities in the Industrial Revolution, and with the coming of the railways the importance of geology was magnified. No one wanted to tunnel through solid granite, or to lay a track across land that would start immediately to subside. There were two main theories about the earth and its geological past. On the one hand, there were the so-called "catastrophists" (like almost every other scientific term of the day, this was coined by Whewell). They believed that every now and then in earth history there had been massive upheavals of a kind not now experienced, and that these had created the mountains and valleys and rivers and seas that we have around us today. Probably these upheavals were not themselves miraculous – that is, events outside the ordinary lawbound course of nature – although the general opinion was that such events resulted in the creation of new species of organisms and this process was surely non-natural. On the other hand, there were what Whewell labeled the "uniformitarians." Represented most importantly by the Scottish-born lawyer turned geologist Charles Lyell, they argued that the ordinary everyday processes of nature – rain, snow, freezing, warming, deposition, erosion, earthquakes, volcanoes, and more – could do everything (Rudwick 1969). Everything, that is, if there were a virtually infinite bank of time on which nature could draw repeatedly. In his *Principles of Geology*, the first volume of which appeared in 1830 and the other two in the years succeeding, Lyell argued for an entirely "actualistic" position (as we now call it): there is nothing in the past that was made by processes that do not still occur in the present and at the same intensities. About organisms, Lyell was somewhat more ambiguous, but overall the reader's impression was that (with the exception of humans) their appearance and disappearance was likewise entirely natural, no special, miraculous interventions being necessary. Ambiguity increases because, as we shall see in a moment, whatever the origins of organisms, apparently they were not evolutionary.

Before he left on the *Beagle* voyage, Darwin enjoyed a crash course on geology with Sedgwick, a leading catastrophist. But he took with him the first volume of Lyell's *Principles* (the later volumes being sent out to him), and at once became a total convert to uniformitarianism. This led to Darwin himself doing a notable piece of scientific theorizing. Among the fascinating phenomena that one finds in tropical waters are the rings of coral that surround islands, or sometimes indeed only the rings, with no islands in the center. Lyell had suggested that perhaps these were the rims of now extinct volcanoes, just breaking the surface of the sea. Darwin pointed out how improbable this must be – so many volcanoes and just at the right height – and argued instead that, since coral can grow only at the surface of the sea, perhaps the islands were sinking and the coral kept growing up to stay at the same (surface) level. Even where there were now no islands, there had once been such lands poking above the waters (Figure 1.1).

Today, general opinion is that Darwin was right. For us, however, the coral reef theory is more than the first fruits of Darwin's creative thinking. It shows how firmly Darwin's worldview was embedded in the Lyellian system. Underlying the catastrophist/uniformitarian

Section of coral reef

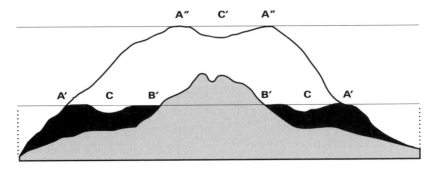

A'A'   Outer edges of the barrier reef at the level of the sea
        with islets on it.
B'B'   The shores of the included island.
CC   The lagoon channel.
A"A"   Outer edges of the reef, now converted into an atoll.
C'   The lagoon of the new atoll.
        NB According to the true scale, the depths of the
        lagoon channel are much exaggerated.

Figure 1.1 Darwin's illustration of his coral reef theory (redrawn from Charles Darwin, *The Structure and Distribution of Coral Reefs*. London: Smith, Elder and Company, 1842, p. 100)

debate was a difference about the putative direction of our planet's history. The catastrophists generally saw the history of the earth as being directional, from hotter to cooler. They tied this direction in with the different organisms that are revealed in the fossil record, arguing that only now was our globe fit to support humankind and the other extant plants and animals. Lyell, to the contrary, saw the earth in a kind of steady state. There may be some fluctuations, but, as with a sine curve, the changes are always within fixed limits. How then could one account for such changes as there were – notably (what the catastrophists held up as their prime piece of evidence) the fact that the fossil plants around Paris suggest that in the past the area was very much warmer than it is today? To speak to this, Lyell introduced what he called his "grand theory of climate" – that the

Charles Darwin

relative temperatures of land and sea are not, as one might think, a function of the distance from the equator, but more of the overall distributions of land and sea around the globe. As these change, so the climates change. Using as his primary evidence the warming effects on Britain of the Gulf Stream, Lyell argued that limited changes occur because the earth is a bit like (to use a modern metaphor) a water bed. Deposition in one area causes the underlying land to sink. This is matched by rising in other areas, thanks to erosion.

Darwin's coral reef theory was a perfect exemplar of the theory of climate. The land beneath the islands was sinking because of the ever increasing buildup of coral. Elsewhere the land must be rising. As it happens, just after the *Beagle* voyage Darwin thought that he had found just such an instance in the Scottish highlands. There is a small valley, Glen Roy, with parallel tracks or roads around its sides. Reasoning that these are the beaches from now vanished water, Darwin argued that once the sea ran into the glen, but since then the land has risen and the sea has vanished – the rising land being a counterpoint to the sinking islands. A nice solution, albeit false: Darwin was quite mistaken. In the last Ice Age, the entrance to the glen was blocked by a glacier. When it melted, the lake that had accumulated behind it ran out.

Darwin the evolutionist

With hindsight, we can see that Darwin's commitment to the climate theory was truly crucial. It explains the sorts of things that he was directed to study. Any scientist will tell you that the answers are easy. The hard part is finding the questions. Following Lyell, Darwin became fascinated with the distributions of organisms, both through time as shown in the fossil record, and through space as shown by geographical distributions. Although he was opaque about the actual origins of organisms, Lyell thought that new organisms tend to be much like those just a little older. Hence, by looking at successive layers in the fossil record one can get ideas about whether or not the land has risen, and by looking at living organisms today one can get ideas about how the geology has changed over the ages. If, say, the animals on either side of a mountain range are very similar, then this suggests that the range is fairly new.

Thus primed, the crucial event for evolutionary thinking came when, in 1835, the *Beagle* visited the group of islands on the equator in the Pacific, the Galapagos archipelago. It is not easy to say just how much the whole question of evolution was now of interest to Darwin. He certainly knew about evolutionary ideas. He had read *Zoonomia*, his grandfather's major prose work on the subject. He had discussed evolution when a teenager at Edinburgh with an enthusiast for the idea, and no doubt the topic was aired at Cambridge by the senior scientists, who were anything but enthusiasts for the idea. Then, in the second volume of the *Principles*, Lyell had discussed the evolutionary ideas of the French biologist Jean-Baptiste Lamarck (1809). Lyell's verdict was negative, leading one to suspect that his silence on the positive aspects of organism appearance was truly a function of total ignorance. Nevertheless, although Lyell himself did not sign up to the Frenchman's ideas, his review was sufficiently thorough that more than one person was converted to evolutionism by the discussion. And Lyell's own ambiguity on the topic of organic origins was certainly enough to stimulate the thinking of a bright young disciple. Indeed, piqued by the less-than-adequate musings of Lyell, the philosopher and astronomer John F. W. Herschel shortly afterwards described the whole question of the origins of organisms as the "mystery of mysteries" – a phrase that Darwin was to use later to preface his own public announcements on the topic (Cannon 1961).

So we know that, as the *Beagle* voyage was coming to an end, Darwin was at least aware of these issues. However, there was to be no immediate road-to-Damascus experience. The Galapagos Islands are very close together, most just a few miles apart. As was his custom, Darwin started making collections of the life found there, particularly the little birds (finches and mockingbirds) on the various islands. There were clearly different species, but it never dawned on Darwin that it might matter which islands the birds occupied. After all, he had just spent several years in South America (Darwin spent much time on land as the *Beagle* worked on charting the coast) and he had seen that, for some species, their members are often in very different and distant habitats. Then the sun did rise and shine. By dining with the governor of the Galapagos, Darwin began to recognize the island differences. Famously, the Galapagos are home to giant tortoises and Darwin learnt that there are recognizably different species on different islands (Figure 1.2). The same was almost certainly true of the

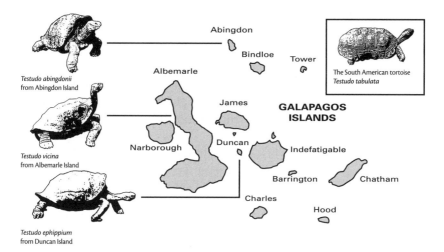

Figure 1.2 Distribution of Galapagos tortoises

birds, and to a Lyellian geologist this surely had to mean something. Biogeography does not happen by chance, and it was clear to Darwin that the denizens of the Galapagos were very similar to (although different from) the denizens of the mainland.

Something significant was at issue, but still Darwin hesitated to hypothesize. The big conceptual move was made in the spring of 1837. Darwin was back home looking at, and trying to make sense of, his collections. He realized that the only sensible answer to the island differences was that original animals had come from the mainland, and once on the Galapagos had changed and evolved after they moved from island to island. Charles Darwin moved across the divide, never again to think that species are fixed, for ever to think that life is in flux. Yet, at the same time, he knew that such a belief would not be viewed with favor by the older members of his group. Hence, Darwin kept very quiet about his thinking, even as he now opened the notebooks in which he began to speculate.

Being a graduate of Cambridge University was a significant factor at this point. Isaac Newton was the university's most famous scientist, and he was a model for all who followed. Most particularly, he was a model for those who aimed to be honored in the halls of science. Newton showed the way. His great achievement was to provide a causal solution to the new physics: his force of gravitational attraction tied together the cosmological speculations of Copernicus and Kepler with the terrestrial mechanics of Galileo. All was now seen to be part of the same system. An ambitious young scientist who wanted to do the same in the life sciences – to be the "Newton of the blade of grass" as Kant had called him (1790; Kant 1928), while denying that such a person could ever exist – had to come up with a causal solution to the problem of organic origins, had to put a Newtonian force behind evolution.

This was Charles Darwin's project until the end of September 1838, when finally he came to the solution. Early on, thanks to his agricultural connections (Shrewsbury lies right in the heart of farming Britain), Darwin realized that the key to changing the forms of animals and plants is systematic selection. The breeder takes his best stock and uses these, and only these, as the parents of the next generation. Rapidly, one gets shaggier sheep and beefier cows and redder and bigger strawberries. What Darwin could not see was how something like this could ever occur in nature. Then – and it is this sort of thing that bears out what I said about Darwin using the ideas that were fed to him – he read a rather conservative socio-political tract by the Anglican clergyman Thomas Robert Malthus. In his *Essay on a Principle of Population* (6th edition, 1826), the first version of which had appeared at the end of the eighteenth century, Malthus was concerned to oppose what he thought were prevalent and unsupported views about the onward progress of humans and their civilizations. Malthus's was a much darker view of life, at the same time one trying to see how God had arranged it that we humans do anything at all. Why do we not just sit around doing nothing? The answer came in a famous deduction. Space and food supplies can at best be increased arithmetically $(1, 2, 3, 4, \ldots)$, whereas, unchecked, population increases geometrically $(1, 2, 4, 8, \ldots)$. There will be inevitable clashes – what Malthus called "struggles for existence"; and grand schemes to alleviate poverty and so forth are bound to end in failure – if anything, they will only make things worse for the next generation. In later versions of his *Essay*, Malthus allowed that such struggles might be avoided if we exercise what he primly called "prudential restraint." I am not sure that he ever truly believed this to be possible in real life. Although his name today tends to be linked to the need for contraception, Malthus himself recoiled in horror at such a disgusting practice.

Darwin seized on the Malthusian ratios, and had there, in his hand, the force behind a natural form of selection. More organisms are born than can survive and reproduce. There is natural variation in populations in the wild. The successful in the struggle (what came to be known as the "fit") will be different from the unsuccessful, and on average and in general the success will be a function of the different characteristics: the successful will be a bit better camouflaged than the unsuccessful, or a bit stronger, or able to go with less food and water, or whatever. Over time this will lead to full-blown change; change, moreover, of a particular kind, namely that which makes for "adaptations" to organs like the hand and the eye that help their possessors in the struggle to survive and reproduce. So we have the biological equivalent to Newtonian gravitation – natural selection or (as it later was called) the survival of the fittest.

The Origin of Species

The private Darwin worked on these ideas for the next five or six years, in 1842 writing up a 35-page sketch of his thinking and in 1844 a 230-page essay (Darwin and Wallace 1958). These remained hidden from view. The public Darwin was now becoming known as a very good young scientist, and also as a travel writer. Darwin's journal from the *Beagle* voyage was published, and quickly made him a known figure in the young Victorian era. (Queen Victoria came to the throne in 1837, and remained there until 1901.) For reasons that are still not fully understood, Darwin started now to show symptoms of an illness that reduced him from the vibrant young adventurer of his youth to the invalid that he was to become and remain for the rest of his life – indigestion, headaches, boils, bad breath, and flatulence, among others. Perhaps it was something physical, possibly a disease picked up in South America; perhaps it was psychological, the stress of his work and his ideas; perhaps it was a consequence of the many appalling potions with which the Victorians dosed themselves, the mildest of which were drugs that today would earn you a long prison sentence. Whatever it was, Darwin became the victim of his ailments. Early in 1839 he married his first cousin, Emma Wedgwood, and with the large settlements that they both received he bought a house in Kent not far from London, where he and Emma set about having and raising a large family. There were ten children, seven of whom lived to maturity.

Darwin became a near recluse, though, as his biographers often note, a recluse who used his illnesses as an excuse to avoid burdensome meetings and other duties. Over the years, as it suited him, he kept in touch with the scientific community, and indeed made new friends among the younger members. Lyell (older than Darwin) persisted as a friend, as did Henslow. Joseph Hooker, a botanist and son of Sir William Hooker, the superintendent in charge of the Royal Botanic Gardens at Kew, became a good friend, and then somewhat later, in the 1850s, the young morphologist Thomas Henry Huxley joined the circle. It is clear that people did truly love Darwin, for in person he was warm and friendly and no doubt genuinely so. However, he used his friends and many, many correspondents as his eyes and ears, to do much of the groundwork of science for him, particularly information collecting, as he turned into a truly obsessive worker, laboring without a break, save only for periods of inactivity brought on by illness.

For reasons that are still not quite clear, Darwin repeatedly postponed the publication of his thinking about evolution. A major factor undoubtedly was that in 1844 an anonymous author – we now know him to have been Robert Chambers, a Scottish publisher – published an evolutionary tome, *Vestiges of the Natural History of Creation*. Crude on science, bold on speculation, it caught the public imagination and drew the excoriation of the Cambridge scientific community, especially Sedgwick and Whewell. Their brightest student was not about to stir up that pot publicly. So, Darwin finished writing books on the geology of the *Beagle* voyage and then in the mid-1840s turned to what was to become an eight-year obsession with barnacle classification – an obsession pursued by dissecting the rather smelly invertebrates sent to him from all over the world, and which led to the publication of massive works on the living species and more books on the fossil representatives.

Finally, in the 1850s Darwin turned back to the question of organic change, and began writing a huge work on evolution. This was interrupted in the summer of 1858 by the arrival of an essay from a young collector and naturalist, then in the East Indies. Alfred Russel Wallace had hit upon virtually the same ideas as Charles Darwin had found some twenty years before. Quickly, Lyell and Hooker arranged for the publication, by the Linnean Society of London, of Wallace's essay together with selected pieces from earlier (hitherto private) writings by Darwin. Then Darwin sat down and in fifteen months wrote up what was published towards the end of 1859 as his definitive statement on the subject: *On the Origin of Species by Means of Natural Selection, or, The Preservation of Favoured Races in the Struggle for Life*. After a long, long delay, Darwin's theory of evolution was there for all to see. An oft-noted but not terribly important point is that Darwin never used the word "evolution" in the *Origin*. "Evolution" was a term only then coming into use to denote the change of species through time, and its use was confined mainly to the change of the individual embryo as it developed. Many used the term "transformation" to mean what we now mean by "evolution." Darwin generally wrote of "descent with modification," although as it happens the last word of the book is "evolved."

Reception

We all know that there was a huge row after the *Origin* was published. But, truly, how successful or unsuccessful was Darwin? This much can be said with near certainty. Very quickly, respectable opinion in Britain and elsewhere in the world (Europe and much of the northern USA, as well as the British Empire) came to accept the evolutionary theory that all organisms, living and dead, are the end results of a long, slow process of law-bound change. This even applies to humans, although most commentators jumped in to argue that souls or spirits require interventions from above. It also seems true that evolution was accepted by most segments of society. Benjamin Disraeli, soon to become the conservative prime minister of Britain, jokingly protested that he was on the side of the angels against evolutionists, but the middle classes and the thinking members of the working class (and in those days there were many such people) embraced organic change. Moreover, although religious people tended more to caution on these matters, in that quarter also there was considerable acceptance of evolution.

It is always hard to justify generalizations of this kind and there have been many exceptions (especially in the American South), but scholars have done very extensive surveys of the literature and especially of the popular journals, newspapers, and magazines (Ellegård 1958). It seems well borne out that acceptance of evolution came quickly. Charles Darwin was a high-profile, well-respected man: his *Beagle* book was a Victorian standard, his barnacle work had turned him almost into a parody of the scientist who spends much time learning more and more about less and less, and his position as a man of solid status (large family, faithful wife, devoted servants, friend of the clergy and squires of the neighborhood, magistrate who judged poachers and other transgressors), who struggles on despite crippling illness, had earned him the affection of his countrymen. It was easy, it was comfortable, to agree with Darwin. He represented all of the things his generation thought praiseworthy. Like the little boy who cried out "But the emperor has no clothes," as soon as Darwin cried out "But evolution is true," within ten years or so almost everyone agreed. And those who disagreed took care to moderate their objections. My favorite piece of confirmatory evidence is the speed with which the examinations at English universities changed from demanding discussion about evolution as such to discussion about the causes.

For here indeed there was continuing controversy (Hull 1973). Most people agreed that natural selection could cause some change. Very few agreed that natural selection could cause all change. There were some serious scientific issues at stake here. In the first place, and most troubling, Darwin had no decent theory of heredity – what we today would call genetics. He needed two things for selection to work properly: a constant source of new variation and a way in which such variation can be transferred from one generation to the next. From detailed study of populations in the wild, not the least of which were his barnacle species, Darwin was convinced that such variation does occur, but he had no real theory to account for why it occurs. Moreover he had no real theory of how such variation gets passed on. A real stumbling block seemed to be that in each generation variations tend to get blended and watered down – a black man and a white woman have a brown child – and it seems that in but a few generations any new feature, however valuable in the struggle for existence, is swamped and wiped out. Darwin certainly knew of features that were not blended and wiped out – sexual features, most obviously. In the mid-1860s he proposed a theory of heredity, "pangenesis," showing how physical features could be preserved, proposing that little gemmules are given off all over the body and collect in the sex cells. Moreover, in later editions of the *Origin* Darwin stepped up the inheritance of acquired characteristics (so-called Lamarckism) as a source of new variation. But ultimately he never really got on top of this problem.

Darwin's other major scientific problem concerned the age of the earth (Burchfield 1974). Darwin never truly specified just how old he thought the earth must be, although in the first edition of the *Origin* he did try his hand at making estimates about the time it had taken for erosion to occur in the part of England where he lived. What he and everyone knew was that a great deal of time had been needed for so slow a mechanism as natural selection to be effective. And this the physicists of his day refused to allow. By working from such factors as the radiation from the sun and the saltiness of the sea, the popular estimate of the earth's age was around a hundred million years, dating from the very beginning of time, when everything was molten and hence far too hot to sustain life. Darwin wriggled as best he could on this question, but here too he recognized that there was a major problem – as, of course, we now know that there was, though it was not Darwin's. The physicists were ignorant of radioactive decay and its warming effect. Now it is thought that the earth is 4.5 (American) billion years old, and that for about 3.75 billion years it has been sustaining life – quite long enough for even a slow process like natural selection.

What did people suggest instead of selection? They were all over the place. Some went one better than Darwin and made the inheritance of acquired characteristics the main driving force behind change. Some, like Thomas Henry Huxley, supposed that occasionally there is a big jump from one form to another, with no intermediates. (This is known as "saltationism," from the Latin *saltus*, "a jump.") Yet others thought that perhaps there is a kind of inner momentum to evolution and that, rather as in embryology, once a species gets going it develops its own internal forces that drive it into new directions and forms. There was one drawback that many of these rival proposals had in common: if you were an active scientist looking for a tool of research, they were not awfully useful. Natural selection was significantly different, for all that most did not take it up: one could apply selection to actual problems, trying to crack them. One man in fact did just this. Henry Walter Bates (1862; Bates 1977), a traveling companion of Wallace, became fascinated with animal mimicry, particularly the way that Amazonian butterflies that were not poisonous would mimic those that were. He argued that selection was responsible, and performed simple but effective experiments to show this. But, with the possible exception of Lamarckism (the theory of inheritance of acquired characteristics), other suggestions simply did not lend themselves much to experiment or test. Major jumps might have happened, but when and how and why and what? Momentum might be important, but what was it and how did it function and was it always in the direction of adaptation?

Evolution as secular religion

The plain fact is that, after Darwin, there simply was not a great deal of interest in causal questions. A lot of people were keen on uncovering the history of life. Although by the time of the *Origin* the main outlines of the fossil record had been established, it was in the decades after the *Origin* that the record was really opened up, especially in America, which poured forth hosts of beasts from the past – Amphicoelous, Allosaurus, Ceratosaurus, Camarasaurus, Diplodocus, Camptosorus, Brontosaurus, and many, many others. More than this. There were many speculations about past histories, drawn by analogy from embryological development. As goes the individual, so apparently goes the group. The German evolutionist Ernst Haeckel (1866) popularized this kind of reasoning with his so-called "biogenetic" law: ontogeny (the development of the individual) recapitulates phylogeny (the development of the group). But when it came to causes, especially causes as the basis for an ongoing evolutionary research program, interest dropped right through the bottom.

The truth is that even Darwin's greatest supporters – in fact, especially Darwin's greatest supporters, if you think of Huxley – did not want to use evolutionary ideas as the foundation of a professional, university-based, area of research and inquiry (Ruse 2005). They wanted to use some parts of biology in this way, for this was the very time when (in Britain and then somewhat later in America) the whole profession of science was being organized and made into a career opportunity for a bright young scholar. This was tied in with more general changes in the ways that the countries were run – toward a world where merit and education rather than simply inherited wealth and status would be the deciding factors in a person's success. These organizers knew full well that if they were to achieve their aims, then they had to show their fellow citizens that what they had to offer were goods that were generally desirable. Organization and reform had to point to future benefits and payoffs. Huxley and friends were very successful at this. Physiology they sold to the medical profession, offering to produce well-grounded biologists who could then go on to medical training, furnished with a solid background in basic science. Anatomy they sold to the teaching profession, arguing that hands-on experience of cutting up fish and rabbits would be better training for real life than rote learning of Latin and Greek. Huxley himself offered summer schools for teachers. His most famous pupil was the novelist H. G. Wells.

Evolution did not make these sorts of promises. It did not cure a pain in the belly and it seemed too risqué for straight, schoolroom teaching. However, there was one use to which it could be put, namely as a kind of ideology – secular religion, if you like – that could substitute in the minds of the new men and women for the old superstitions (otherwise known as Christianity) of the past. Rightly, Huxley saw the Church of England as allied with all of the conservative forces in Britain against which he and other reformers were battling, and evolution fit the role of a kind of popular science or world picture that could replace it. Like Christianity, evolution told of origins, it told of humankind and (in the opinion of Huxley and friends) it put us not only last-appearing but at the top, and for many it offered a kind of up-to-date version of the Sermon on the Mount. We will learn more later about "social Darwinism," but in essence it directed people to do whatever will best further the survival and continued success of the human species, most particularly that fragment of the human species to which this kind of evolutionary enthusiast belonged. In short, far from being a vigorous new branch of science – which had surely been Darwin's aim when he wrote the *Origin of Species* – evolutionary theorizing became a part of the social fabric of forward-looking Britain (and America and elsewhere). It was popular, but with hindsight not necessarily for the right reasons or for the right purposes.

What of Darwin himself after the *Origin*? For the two decades left to him, he kept working hard. He wrote books on a variety of topics, from orchids to climbing plants, from agriculture to earthworms. His main work, however, was on our own species, *Homo sapiens*. Darwin had never doubted that we humans are part of the world of life. We must have evolved and for essentially the same reasons as other animals. HMS *Beagle* carried, and returned to the tip of South America (Tierra del Fuego), three natives, who had been taken on a previous voyage and educated in England. Within a very short time, these three, who had apparently acquired a solid veneer of British culture, reverted to their original state of complete savagery – at least as judged by the young ship's naturalist. Darwin learnt a lesson that he never forgot. The line dividing the most refined of humans and the most degraded of brutes is very fine indeed. It is remarkable how, at a time when everyone else was jumping around and arguing over the question of the status of humans – going against his own philosophy, Lyell could never quite bring himself to think that human origins were entirely natural – Darwin was entirely cold-blooded on the matter. We are animals. End of argument. Most significantly, the very first (private notebook) jottings that we have dealing unambiguously with natural selection, from late in 1838, apply the mechanism to humans – and to their mental abilities, into the bargain.

In the *Origin*, wanting first to get his basic theory on the table as it were, Darwin merely made very brief reference to humankind, so that no one could think he was dodging the issue. "In the distant future I see open fields for far more important researches. Psychology will be based on a new foundation, that of the necessary acquirement of each mental power and capacity by gradation. Light will be thrown on the origin of man and his history" (Darwin 1859, 488). That was it. For the time being at least. Towards the end of the 1860s, Darwin returned to the issue of humankind, and wrote a book where the focus was essentially on our species (followed by another, almost a supplement, on the emotions). Picking it up and looking at it today, the "human" book, *The Descent of Man* (1871), is rather odd. It may be essentially on humans. It is not primarily on humans. From the beginning, Darwin had always believed that, along with natural selection, there is a secondary form of selection, which he called "sexual selection," meaning the selection that comes from competition for mates. He divided it into two kinds. First there is sexual selection through male combat, as when two stags fight for the female and in successive generations the stags' antlers get bigger and bigger. Second there is sexual selection through female choice, as when females choose between males with which to mate – for instance, a peahen deciding between two displaying peacocks. Undoubtedly these two kinds come from the world of breeders: natural selection is like choosing bigger and fleshier cattle, sexual selection through male combat is like two fighting cocks going at each other, and sexual selection through female choice is like choosing the dog that most closely fits the standards that one favors. However, although sexual selection was no mere add-on, but something that came right from the heart of Darwin's thinking, for much of his life he made little of it.

Then when he turned full time to humans, Darwin began to think that sexual selection might be an important mechanism in its own right. Wallace had gone somewhat off the track. By the late 1860s, he had embraced spiritualism and, thinking that unseen forces must have been responsible for human evolution, he was denying that natural selection could account for human intelligence (Wallace 1905). He argued that many human features, not just intelligence but things like the lack of hair, are beyond causes as we know them. Fighting something of a rearguard action, Darwin decided that Wallace was right about natural selection as such, but wrong about selection generally. Sexual selection could pick up the slack. In particular, in humans, males compete for females and females choose the males they like. This leads to all sorts of racial and sexual differences in their offspring, as well as to such things as improved intelligence and so forth. To make this case, Darwin went off in the middle of *The Descent of Man* into a long digression about sexual selection in the animal world generally, and only towards the end did he return to our species to draw his conclusions – conclusions that left him where he had come in almost forty years before: humans are animals and as such we have evolved like every other living thing. No exceptions.

Work to be done

We draw to the end of the story of Charles Darwin the person. From the viewpoint of evolutionary theory, there was yet a long way to go. Scientifically, at the beginning of the twentieth century, the age-of-the-earth question cleared itself up quickly when it was realized that radioactive decay and the warmth it generates makes for a much longer earth history. Heredity, building on the insights of Gregor Mendel (of whom more later), took much more work and time. Even when its study started to pick up pace, there were barriers to be overcome. Early geneticists, to use the name given to those who study variation, tended naturally to work with major variations, and hence thought that these must be the key to significant evolutionary change. It took some time for this kind of saltationism to subside and for selection and genetics to work together rather than apart.

Indeed, it was not until the 1930s that the real synthesis came – appropriately the Darwin–Mendel conjunction was known (in Britain) as "neo-Darwinism" or (in America) as the "synthetic theory of evolution" – and from then evolution had its working theory (Mayr and Provine 1980; Ruse 1996). At the same time, thanks to its now professional practitioners, it began to distance itself from its role as something purely at the level of popular science or secular religion. With selection backed by Mendelism (that is, genetics) now in place, it was possible for full-time scientists to work on evolutionary problems, experimenting, observing, and hypothesizing, just like scientists in other fields. We shall talk more on these matters in later chapters. Now is our time to take leave of the individual. Darwin grew old, loved by his family and friends, greatly respected by his countrymen and by many in other lands. When he died, there was little debate about what absolutely had to happen. He must be placed in that Valhalla of English heroes, Westminster Abbey. And there he still lies, right next to Isaac Newton, who led the way, doing in physics what Darwin was to do in biology.

2

On the Origin of Species

On the Origin of Species appeared at the end of November, 1859. Darwin had been worried about its potential sales, but his canny Scottish editor, John Murray, was more confident and printed 1,250 copies. It is often said that they sold out on the first day. This is not quite true. Murray had twice-yearly promotions to which he invited booksellers. It was they who ordered all of the copies of the *Origin*, prompting Murray to ask Darwin to start at once on a second edition. In all there were six editions, the last appearing in 1872 and very much revised and augmented as Darwin struggled to keep up with objections and criticisms. Following general practice today, I shall focus on the first edition, mentioning revisions only as necessary. As we can see with hindsight, too often Darwin messed with his text unnecessarily, answering objections many of which were misguided. It would have been better, and certainly easier on the reader, if he had not kept tinkering with passages until they became long and convoluted.

This chapter will be expository. The next chapter will take up the analysis.

Artificial selection

Darwin started the *Origin* by turning to the world of the breeders of animals and plants. The seventy-five years or so leading up to the mid-nineteenth century had been a time of incredible advances in this area. An Industrial Revolution demands an Agricultural Revolution, as far fewer farm workers must feed many more city dwellers. Appreciation of the efforts made to increase agricultural production had been a

Figure 2.1 Varieties of fancy pigeon, originating in the rock pigeon of India: (*left to right, top row*) baldheads, pouters (with stuffed shirt), Jacobins (with head ruff), magpie, and swallow; (*left to right, bottom row*) fantails, with carrier between, Brunswick, nun, and turbit

crucial step in Darwin's journey to the theory of natural selection. Now he used it to lead the reader to the same conclusions. Wanting to gain some hands-on experience of the whole topic, Darwin himself had taken up the breeding of pigeons, even to the extent of joining clubs for pigeon fanciers. In his opening chapter, turning his knowledge to full advantage, Darwin showed how, despite the many different varieties of pigeon and their significant defining character-istics, all the evidence points to the birds having a common origin in the rock pigeon of India (Figure 2.1): a cameo of diversification from a single earlier form.

> Great as the differences are between the breeds of pigeons, I am fully con-vinced that the common opinion of naturalists is correct, namely, that all have descended from the rock-pigeon (Columba livia), including under this term several geographical races or sub-species, which differ from each other in the most trifling respects.
>
> (Darwin 1859, 23)

But how do breeders, whether those pursuing pleasure like the pigeon fanciers or those pursuing profit like the farmers, get their changes? Really remarkable things are possible and it all comes about through the selection of the better or more desirable forms. You pick the organisms with the features you want and you breed from these and these alone.

> Breeders habitually speak of an animal's organisation as something quite plastic, which they can model almost as they please. If I had space I could quote numerous passages to this effect from highly competent authorities . . . Lord Somerville, speaking of what breeders have done for sheep, says: 'It would seem as if they had chalked out upon a wall a form perfect in itself, and then had given it existence.' That most skilful breeder, Sir John Sebright, used to say, with respect to pigeons, that 'he would produce any given feather in three years, but it would take him six years to obtain head and beak.'
>
> (p. 31)

Darwin concluded his discussion of artificial selection by referring to something that he called "unconscious selection." Often breeders bring on changes that they did not really intend. Comparing, say, a line of dogs from one century to the next, or two groups of animals that different breeders have created from the same original stock, almost invariably one sees significant, albeit unintended, variations. Subtle changes have occurred – under the radar, as it were: breeders have kept their eyes on the main targets, but unplanned shifts have frequently occurred. Perhaps even more than active artificial selection, this phenomenon of change without direct intention prepared the way for an understanding of natural selection. Here was change brought on blindly, without any aim of achieving some pre-specified goal, just like the situation in the wild.

His observations on unconscious selection underline an important point about the significance of this whole discussion of artificial selection for Darwin. One might think that by talking about breeding and the power of artificial selection, he was simply leading the reader in gently, relaying his own route to discovery, and preparing what is to come next. This is certainly true, but it is not the whole story. Darwin was also preparing to use breeding and artificial selection as evidential support for his overall theory of evolution through selection. What happens in the farmyard is analogous to what happens in nature, and because of the first we should be prepared to accept the second. As the *Origin* proceeds, again and again Darwin referred back to artificial selection to illustrate and support points he made about what happens in the wild.

The struggle for existence

Darwin next set up and explained his interpretation of the main mechanism of natural selection. First he reported on the widely documented existence of natural variation in wild populations. Without this there could be no overall ongoing change. At the same time, he downgraded the significance of species, often then (and now) regarded as the truly significant group in nature – that which, in some sense, is not simply read into nature by us but exists objectively in the living world.

> I look at the term species, as one arbitrarily given for the sake of convenience to a set of individuals closely resembling each other, and that it does not essentially differ from the term variety, which is given to less distinct and more fluctuating forms. The term variety, again, in comparison with mere individual differences, is also applied arbitrarily, and for mere convenience sake.
>
> (p. 52)

After this preliminary discussion, Darwin was ready to plunge into the central ideas. First there was the struggle for existence. This is a major plank underpinning natural selection, and Darwin had to spend some time discussing the idea, preventing misconceptions, and convincing the reader just how significant and widespread a phenomenon it truly is. You look around at nature, and more often than not it seems a happy, harmoniously balanced system – birds sing, butterflies flutter, trees blow in the wind, and mammals stand gratefully in the shade enjoying God's creation. But a darker picture lies close to the surface. Nothing lasts long without blood and violence, or what substitutes for this in the plant world. Birds eat butterflies and are themselves subject to predation, starvation, and worse. Trees compete for space, when they are not being grazed out of existence by voracious mammals. Animals fight among themselves and are constantly looking for prey or fearing that they themselves will be preyed upon. It's a jungle out there and this is true of even the prettiest and quietest of English landscapes. Never let your guard down for a second, or you may never again have a guard to raise.

How can this be? It is simply a result of the Malthusian pressures.

A struggle for existence inevitably follows from the high rate at which all organic beings tend to increase. Every being, which during its natural life-time produces several eggs or seeds, must suffer destruction during some period of its life, and during some season or occasional year, otherwise, on the principle of geometrical increase, its numbers would quickly become so inordinately great that no country could support the product. Hence, as more individuals are produced than can possibly survive, there must in every case be a struggle for existence, either one individual with another of the same species, or with the individuals of distinct species, or with the physical conditions of life. It is the doctrine of Malthus applied with mani-fold force to the whole animal and vegetable kingdoms; for in this case there can be no artificial increase of food, and no prudential restraint from marriage. Although some species may be now increasing, more or less rapidly, in numbers, all cannot do so, for the world would not hold them.

(pp. 63–4)

There is ongoing conflict, although as Darwin was keen to stress, this conflict is not necessarily always of a direct, violent, and limbs-torn-and-destroyed kind. It could be more nuanced.

I should premise that I use the term Struggle for Existence in a large and metaphorical sense, including dependence of one being on another, and including (which is more important) not only the life of the individual, but success in leaving progeny. Two canine animals in a time of dearth, may be truly said to struggle with each other which shall get food and live. But a plant on the edge of a desert is said to struggle for life against the drought, though more properly it should be said to be dependent on the moisture. A plant which annually produces a thousand seeds, of which on an aver-age only one comes to maturity, may be more truly said to struggle with the plants of the same and other kinds which already clothe the ground.

(pp. 62–3)

It is worth noting that, for Darwin, the relationships between organisms existing at the same time in some shared geographical area – that part of biological science that we would call "ecology" (the term was invented by Ernst Haeckel in 1866) – were very much part and parcel of his thinking about nature generally. He realized that the struggle could involve chain-like relationships between very different organisms, and that balance could result, until or unless something external came in to disturb things. In a famous illustration, he pointed out that the clover in an agricultural area needs fertilization by bumble bees (Darwin calls them "humble bees"), and these bees are predated by field mice. However, in turn the mice fall victim to marauding cats, and so we have a case where the success of a flower is a function of the number of a certain species of pet mammal. Take one link out of the chain and all collapses. For a hundred and fifty years, jokers have pointed out that Darwin missed one final link, namely the owners of the cats. Take the old ladies out of a village and the mice increase, the bees decrease, and soon the clover is gone.

Natural selection

More seriously, the point is that nature is a complex of interacting relationships, and this of course applies particularly to the chief mechanism of change. (Darwin never denied subsidiary mechanisms, notably the Lamarckian inheritance of acquired characteristics.) With the case now made for the struggle, natural selection could make its entrance.

> Let it be borne in mind in what an endless number of strange peculiarities our domestic productions, and, in a lesser degree, those under nature, vary; and how strong the hereditary tendency is. Under domestication, it may be truly said that the whole organisation becomes in some degree plastic. Let it be borne in mind how infinitely complex and close-fitting are the mutual relations of all organic beings to each other and to their physical conditions of life. Can it, then, be thought improbable, seeing that variations useful to man have undoubtedly occurred, that other variations useful in some way to each being in the great and complex battle of life, should sometimes occur in the course of thousands of generations? If such do occur, can we doubt (remembering that many more individuals are born than can possibly survive) that individuals having any advantage, however slight, over others, would have the best chance of surviving and of procreating their kind? On the other hand, we may feel sure that any variation in the least degree injurious would be rigidly destroyed. This preservation of favourable variations and the rejection of injurious variations, I call Natural Selection.

(pp. 80–1)

Remember that this quotation, like the others, is from the first edition of the *Origin*. At this point, Darwin did not use the alternative name for natural selection, "survival of the fittest." This term was invented by his fellow English evolutionist Herbert Spencer and urged on Darwin by Wallace as something with fewer anthropomorphic connections. Darwin introduced the synonym only in later editions of the *Origin*.

The already introduced ecological factors made another appearance at this point. Why should one think that natural selection is going on, all of the time, in virtually every dimension? Why should one think that altering one factor in the web of nature is going to have repercussions throughout the biosphere? Simply because organisms are so tightly interwoven: as soon as one thing changes so does everything else, and selection swings at once into action. "We may conclude, from what we have seen of the intimate and complex manner in which the inhabitants of each country are bound together, that any change in the numerical proportions of some of the inhabitants, independently of the change of climate itself, would most seriously affect many of the others" (p. 81). Sexual selection was also introduced here (Figure 2.2). First male combat:

> this depends, not on a struggle for existence, but on a struggle between the males for possession of the females; the result is not death to the unsuccessful competitor, but few or no offspring. Sexual selection is, therefore, less rigorous than natural selection. Generally, the most vigorous males, those which are best fitted for their places in nature, will leave most progeny. But in many cases, victory will depend not on general vigour, but on having special weapons, confined to the male sex.
>
> (p. 88)

Then female choice:

> Amongst birds, the contest is often of a more peaceful character. All those who have attended to the subject, believe that there is the severest rivalry between the males of many species to attract by singing the females. The rock-thrush of Guiana, birds of paradise, and some others, congregate; and successive males display their gorgeous plumage and perform strange antics before the females, which standing by as spectators, at last choose the most attractive partner. . . . I can see no good reason to doubt that female birds, by selecting, during thousands of generations, the most melodious or beautiful males, according to their standard of beauty, might produce a marked effect.
>
> (pp. 88–9)

Figure 2.2 Sexual selection: *Lophornis ornatus* female and male (from Darwin 1871, 2: 76)

The mechanisms are now on the table, as it were. Selection, natural and sexual, is there for all to see. Note something that is going to be a major theme throughout this book I am writing. Natural selection is a mechanism that does more (or, rather, purports to do more) than simply explain change. It explains change in a particular direction: it explains the distinctive features of organisms. Living beings

On the Origin of Species

are not just thrown together willy-nilly. They work, they function, they have features that help them to survive and reproduce. They have eyes and teeth and ears and noses and leaves and bark and flowers and much more. Organisms adapt and their adaptations or contrivances help in the struggle for existence. This was a crucial insight of Darwin and it must never be underestimated. It will come up again and again and again.

In the light of this, it is hardly odd to suggest that next the reader might have supposed that Darwin would give a list, a catalogue, of examples of natural and sexual selection in action, of cases where adaptation is produced. This does not mean that no evidence has yet been given. Clearly Darwin's account of artificial selection is intended to play that role, in some respects. The shagginess of the sheep, the beefiness of the cow, and the fleshiness of the vegetable are all examples of characteristics analogous to the adaptations of nature. But what about some examples of selection in the wild? The simple fact is that we draw a blank. There is nothing. All we get is an example of how selection might be expected to work. Explicitly acknowledging that he was offering an "imaginary illustration," Darwin wrote:

> Let us take the case of a wolf, which preys on various animals, securing some by craft, some by strength, and some by fleetness; and let us suppose that the fleetest prey, a deer for instance, had from any change in the country increased in numbers, or that other prey had decreased in numbers, during that season of the year when the wolf is hardest pressed for food. I can under such circumstances see no reason to doubt that the swiftest and slimmest wolves would have the best chance of surviving, and so be preserved or selected, provided always that they retained strength to master their prey at this or at some other period of the year, when they might be compelled to prey on other animals.
>
> (p. 90)

Darwin added that the actual way wolves might evolve would depend on the circumstances and the available prey, and that those who go after fast prey might develop features different from those going after slower prey. There might be adaptations for one lifestyle and adaptations for another lifestyle. In fact, there does seem some evidence that this might actually have happened. "I may add, that, according to Mr. Pierce, there are two varieties of the wolf inhabiting the

Catskill Mountains in the United States, one with a light greyhound-like form, which pursues deer, and the other more bulky, with shorter legs, which more frequently attacks the shepherd's flocks" (p. 91). But beyond that Darwin did not care to go, nor are other illustrative examples of selection in action any better supported by empirical evidence. The *Origin of Species* of 1859 simply does not offer this sort of collateral.

What the *Origin* does offer is Darwin trying to tease out the general, more theoretical implications of selection. (From now on, in the *Origin*, this is almost exclusively natural rather than sexual selection.) Readers often complain that, although Darwin's book is called the *Origin of Species*, he did not spend much time on this topic. This is not quite fair, although it is true that it is not the central focus of the book and the discussion is fragmented. But it was on Darwin's mind and he picked up on it as soon as selection was introduced. Interbreeding brings on uniformity. "Intercrossing plays a very important part in nature in keeping the individuals of the same species, or of the same variety, true and uniform in character" (p. 103). Species, however, show differences between their members. How can this be? The Galapagos experience kicks in here. Organisms get isolated and then differences appear and are able to ingratiate themselves because the leveling processes that would normally suppress them are absent.

> Isolation, also, is an important element in the process of natural selection. In a confined or isolated area, if not very large, the organic and inorganic conditions of life will generally be in a great degree uniform; so that natural selection will tend to modify all the individuals of a varying species throughout the area in the same manner in relation to the same conditions. Intercrosses, also, with the individuals of the same species, which otherwise would have inhabited the surrounding and differently circumstanced districts, will be prevented.
>
> (p. 104)

Darwin had more to say on this topic later. He was not convinced that isolation alone is the key. Often, isolated groups are very small and will probably not develop enough internal variation to have major effects. One needs big groups, although Darwin added that here too one may find that barriers divide off subpopulations, in which case isolation could still be important. "Moreover, great areas, though now continuous, owing to oscillations of level, will often have recently existed in a broken condition, so that the good effects of isolation will generally, to a certain extent, have concurred" (p. 106).

The principle of divergence

A crucial corollary to natural selection was what Darwin called his "principle of divergence." Darwin had always thought of evolution as essentially branching, a concept that came from the Galapagos experience, where organisms moved from island to island before they changed. But what were the causes of branching? Was it always simply a byproduct of island hopping, without any real significance in itself, or could it be also subsumed within natural selection? The issue now was not so much isolation, but why isolation might make any difference. Why did groups go different ways? Leaving aside the immediate causes of branching, would there be any ultimate causes? Is there some reason for the multitude of species that we have – so many different forms – and does natural selection play a part in this? Between the early formulations of the theory and the *Origin*, Darwin convinced himself that selection did have a role. Different species could occupy (what we would call) different ecological niches, and those occupying the middle ground would be eliminated by those at the extremes, which are better suited to the particular characteristics of their respective environments.

> The advantage of diversification in the inhabitants of the same region is, in fact, the same as that of the physiological division of labour in the organs of the same individual body a subject so well elucidated by Milne Edwards. No physiologist doubts that a stomach by being adapted to digest vegetable matter alone, or flesh alone, draws most nutriment from these substances. So in the general economy of any land, the more widely and perfectly the animals and plants are diversified for different habits of life, so will a greater number of individuals be capable of there supporting themselves. A set of animals, with their organisation but little diversified, could hardly compete with a set more perfectly diversified in structure.
>
> (pp. 115–16)

The point, simply, is that nature provides different challenges and opportunities – many of them brought on by organisms themselves – and

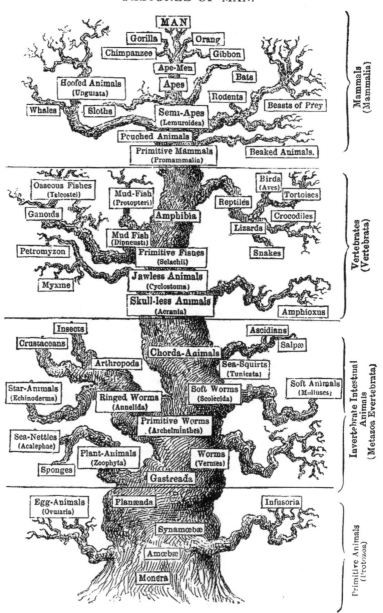

Figure 2.3 The "Tree of Life" as illustrated by Ernst Haeckel (from Ernst Haeckel, *The Evolution of Man.* New York: Appleton, 1896, 2: 188)

On the Origin of Species

hence, under the pressure of selection, life diversifies to grab all of the options available. And with this, Darwin was now ready to present his great metaphor of the "Tree of Life" (Figure 2.3).

> The affinities of all the beings of the same class have sometimes been represented by a great tree. I believe this simile largely speaks the truth. The green and budding twigs may represent existing species; and those produced during each former year may represent the long succession of extinct species. At each period of growth all the growing twigs have tried to branch out on all sides, and to overtop and kill the surrounding twigs and branches, in the same manner as species and groups of species have tried to overmaster other species in the great battle for life. The limbs divided into great branches, and these into lesser and lesser branches, were themselves once, when the tree was small, budding twigs; and this connexion of the former and present buds by ramifying branches may well represent the classification of all extinct and living species in groups subordinate to groups.
>
> (pp. 129–30)

Variation and difficulties

Next came a couple of transitional chapters, as Darwin filled in details before he was ready for the arguments of the rest of the book. One chapter was on variation, where Darwin discussed all that was known about variation and its causes and transmissions. Reading the chapter, one can see that Darwin did not take the topic lightly and, moreover, that he knew a great deal about the subject – normal patterns, exceptions, skipping of generations ("reversion"), correlation, and much more. At the same time, Darwin was moving in the dark, without a coherent or unified theory. For whatever reason, probably the fact that no one seemed much impressed, he never introduced his theory of pangenesis into the later editions of the *Origin*. The other chapter is titled "Difficulties on Theory," and it delivered just this. For a start, Darwin picked away again at the scab of species formation ("speciation"). He really did incline to think that isolation was essential, but he worried that small groups would simply not have the potential to produce the variations needed to make new species. Perhaps more commonly what happens is that a group breaks apart and the middle, linking, group is smaller in some way and hence more likely to be eliminated because it cannot produce the variation of the larger groups.

> I may illustrate what I mean by supposing three varieties of sheep to be kept, one adapted to an extensive mountainous region; a second to a comparatively narrow, hilly tract; and a third to wide plains at the base; and that the inhabitants are all trying with equal steadiness and skill to improve their stocks by selection; the chances in this case will be strongly in favour of the great holders on the mountains or on the plains improving their breeds more quickly than the small holders on the intermediate narrow, hilly tract; and consequently the improved mountain or plain breed will soon take the place of the less improved hill breed; and thus the two breeds, which originally existed in greater numbers, will come into close contact with each other, without the interposition of the supplanted, intermediate hill-variety.
>
> (p. 177)

In the technical language of today, Darwin was wrestling with the problem of whether speciation is always "allopatric" – that is, involving isolation – or whether it could be "sympatric," where groups break away without actual physical separation.

Another problematic topic that Darwin tackled was the supposed impossibility of the evolution of features that are very complex and seem to work perfectly – the eye, most notably. How on earth could something like this have come about through natural selection? Even simply, at the level of evolution, this seems a major difficulty. There is no way that we are going to see such a development represented in the fossil record. Recognizing that the objection is as much psychological as physiological, Darwin's solution was to shift the framework of proof from a sequence through time to a sequence through space. We should remember that in the living world today types of eye range from the most primitive light-sensory mechanisms to the most complex vertebrate eye. "I can see no very great difficulty (not more than in the case of many other structures) in believing that natural selection has converted the simple apparatus of an optic nerve merely coated with pigment and invested by transparent membrane, into an optical instrument as perfect as is possessed by any member of the great Articulate class" (p. 188). Claims about impossibility must first deal with such genuinely observable empirical facts. "If it could be demonstrated that any complex organ existed, which could not possibly have been formed by numerous, successive, slight modifications, my theory would absolutely break down. But I can find out no such case" (p. 189).

I should say that in moving the eye sequence from a range through time to a range through space, Darwin was rather cleverly taking a leaf from the non-evolutionist book. At the back of Darwin's mind, and that of his readers, was a German movement, dating back to the beginning of the nineteenth century, known as *Naturphilosophie*. Promoted by philosophers like Friedrich Schelling and G. W. F. Hegel, by scientists like the morphologist Lorenz Oken, and by the poet, man of letters, and thinker Johann Wolfgang von Goethe, this movement or philosophy saw the whole world, organic and inorganic, as exhibiting fundamental underlying ideas or archetypes. Patterns repeat themselves as the six sides of the snowflake (a favorite example) repeat themselves. The world, especially in the philosophy of Hegel, was seen as improving, progressively, and this was reflected in the history of life. For some this history was physical and evolutionary; for most it was an idea, and physical evolution was belied by the insurmountable gaps in the fossil record. The truth of this life vision is shown by the way that it is repeated in the history of the individual organism, and demonstrated in the progressive spectrum of life today.

"One may consider it as henceforth proved that the embryo of the fish during its numerous families, and the type of fish in its planetary history, exhibit analogous phases through which one may follow the same creative thought like a guiding thread in the study of the connection between organized beings" (Agassiz 1885, 1, 369–70) – so said the eminent ichthyologist and opponent of evolution Louis Agassiz, native of Switzerland and teacher at Harvard, who through his intimate knowledge of glaciers was the author of the Ice Age theory and, incidentally, the man who showed Darwin wrong on Glen Roy.

Evolution aside, this was simply not the world picture of Charles Darwin. Above all else, there is an inevitability, a necessity, to *Naturphilosophie*. Just as the organism develops in the way that it does, so the history of life unfurls in the way that it does (Richards 2003). For Darwin, natural selection is too contingent and relativistic for this. But Darwin was always happy to take up the ideas and premises of *Naturphilosophie*, and here, in comparing the eye through time and the eye through space, he turned *Naturphilosophie* very neatly to his own ends. He did so again in drawing his list of difficulties to an end, when he turned the table on his potential critics by pointing proudly to the fact that his theory solves one of the biggest issues in biology, namely the relationship between what all regarded as the two great features of the organic world – its adaptedness and the interconnections of different organisms. For the *Naturphilosoph*, the interconnections were primary, the real clues to organic understanding. For Darwin, they were important, but secondary to evolution through selection.

> It is generally acknowledged that all organic beings have been formed on two great laws Unity of Type, and the Conditions of Existence. By unity of type is meant that fundamental agreement in structure, which we see in organic beings of the same class, and which is quite independent of their habits of life. On my theory, unity of type is explained by unity of descent. The expression of conditions of existence, so often insisted on by the illustrious Cuvier, is fully embraced by the principle of natural selection. For natural selection acts by either now adapting the varying parts of each being to its organic and inorganic conditions of life; or by having adapted them during long-past periods of time: the adaptations being aided in some cases by use and disuse, being slightly affected by the direct action of the external conditions of life, and being in all cases subjected to the several laws of growth. Hence, in fact, the law of the Conditions of Existence is the higher law; as it includes, through the inheritance of former adaptations, that of Unity of Type.
>
> (p. 206)

Instinct and hybrids

We are now at about the point of the *Origin* where Darwin was swinging from one kind of discussion to another. Thus far he had been establishing the basic theory and its mechanism of natural selection, but now he was about to move on to the application of that theory and mechanism, with a survey of the biological world. Darwin would argue that evolution through natural selection explains the facts of biology, and conversely that these facts justify acceptance of evolution through natural selection. Instinct is the first topic up. Darwin did not even think it worth remarking that what an organism does is as important as what an organism is. There is no point in having the physique of Tarzan if your thoughts and behaviors are exclusively those of a

On the Origin of Species

philosopher. Biologically speaking, the sex-obsessed runt is ahead of the chaste intellectual of perfect proportions. What fascinated Darwin – not least because this was a topic of great general interest when he was writing the *Origin* – was that type of behavior that comes not through thought and deliberation, or through habit, but innately, by nature. Although Darwin was a little dubious about the status of such "instinctive" behavior, he was not at all dubious about the significance of natural selection. The real question is whether instinct is better regarded as a phenomenon that is explained by selection and hence a support for the theory, or as a phenomenon that is explained away by selection and hence merely not a refutation of it – in other words, more a "difficulty on the theory" than anything else. Although he did describe instinct in this second way, for Darwin it was clearly more than this and something supportive of selection as much as something to be explained away. The discussion of instinct also really showed that Darwin was more than simply a fact gatherer – that is, a gatherer of other people's facts – and was a serious researcher in his own right.

The chapter on instinct starts by affirming an observation that was absolutely central to the Darwinian vision: namely, that the basic building blocks of evolution, the variations, are small. Selection works on the almost minute – individual differences, Darwin called them – rather than the large – saltations.

> No complex instinct can possibly be produced through natural selection, except by the slow and gradual accumulation of numerous, slight, yet profitable, variations. Hence, as in the case of corporeal structures, we ought to find in nature, not the actual transitional gradations by which each complex instinct has been acquired for these could be found only in the lineal ancestors of each species but we ought to find in the collateral lines of descent some evidence of such gradations; or we ought at least to be able to show that gradations of some kind are possible; and this we certainly can do.
>
> (p. 210)

Darwin discussed a number of well-known but tricky instances of instinct at work: the cuckoo laying its eggs in the nests of others, the slave-making abilities of certain species of ant, and above all the intricate nest building of the honey bee, which produces combs made from wax in the form of layer upon layer of hexagonal containers, in which eggs are laid and the young reared. How can instincts like these possibly come into being and how in particular can they be produced by selection? Darwin showed that there are other species of bee that produce less intricate and standardized combs and that (as with the eye) it is reasonable to think that through space we see what is presumed to have happened through time. Running various experimental tricks involving colored sheets of wax, Darwin showed how the bees function and how, although they are intricate, one can break down their actions into fairly simple components. And above all, he showed how the honey bees produce combs that maximize the benefits for the efforts expended. Bees need honey to produce the wax. Getting honey requires effort. Hence those bees that most efficiently use their wax are those bees better adapted than others. The honey bee scores above all others. "Beyond this stage of perfection in architecture, natural selection could not lead; for the comb of the hive-bee, as far as we can see, is absolutely perfect in economising wax" (p. 235). This is a protoform of thinking in terms of what we shall later see labeled as an "optimality model."

> Thus, as I believe, the most wonderful of all known instincts, that of the hive-bee, can be explained by natural selection having taken advantage of numerous, successive, slight modifications of simpler instincts; natural selection having by slow degrees, more and more perfectly, led the bees to sweep equal spheres at a given distance from each other in a double layer, and to build up and excavate the wax along the planes of intersection. The bees, of course, no more knowing that they swept their spheres at one particular distance from each other, than they know what are the several angles of the hexagonal prisms and of the basal rhombic plates. The motive power of the process of natural selection having been economy of wax; that individual swarm which wasted least honey in the secretion of wax, having succeeded best, and having transmitted by inheritance its newly acquired economical instinct to new swarms, which in their turn will have had the best chance of succeeding in the struggle for existence.
>
> (p. 235)

Completing the chapter, Darwin turned to a well-known and again often discussed phenomenon, sterility in the social insects – something he proudly noted could not possibly be explained by the Lamarckian inheritance of acquired characteristics. But natural selection could and did step in here. Darwin argued that in such cases it is allowable to consider the whole family as the unit of selection rather than the

On the Origin of Species

individual. So inasmuch as the sterile workers are helping the group, their sterility can be produced by selection.

> How the workers have been rendered sterile is a difficulty; but not much greater than that of any other striking modification of structure; for it can be shown that some insects and other articulate animals in a state of nature occasionally become sterile; and if such insects had been social, and it had been profitable to the community that a number should have been annually born capable of work, but incapable of procreation, I can see no very great difficulty in this being effected by natural selection. . . . selection may be applied to the family, as well as to the individual, and may thus gain the desired end.
>
> (p. 237)

As always, Darwin drew an analogy with the domestic world – in this case how breeders will kill organisms for consumption without breeding from them, but then go back to the family again for fresh stock.

Hybridism was also technically a "difficulty on theory," and was in fact (much more so than instinct) basically a listing of phenomena and of results of breeding rather than a systematic showing of how selection leads to a body of results in one area of biology. Having ignored it to this point, Darwin now grasped straight up front the essential idea that the defining mark of a species is that its members are reproductively isolated from all other organisms. You try to breed a member of one species with a member of another species, and by and large you fail: at best, as with the horse and donkey, you get a sterile hybrid – in this case the mule. How then does this all compare with the status of varieties, which are not separated by such a defining mark? The different varieties of pigeon can all be interbred, for instance. How can one make the claim made earlier that species are, as it were, just varieties writ large? Darwin pointed out that, virtually by definition, if something interbreeds it is a variety and if it does not it is a species. So the intermediates between varieties and species are to an extent defined out of existence. Also he pointed out that here the domestic parallel is not necessarily very helpful. Breeders are simply not interested in promoting infertility or sterility. It is not aimed for, so you should not expect to find it. But even with these points put aside, Darwin was by no means sure that there are no intermediates, cases of varieties on their way to full species-hood.

Gärtner kept during several years a dwarf kind of maize with yellow seeds, and a tall variety with red seeds, growing near each other in his garden; and although these plants have separated sexes, they never naturally crossed. He then fertilized thirteen flowers of the one with the pollen of the other; but only a single head produced any seed, and this one head produced only five grains. Manipulation in this case could not have been injurious, as the plants have separated sexes. No one, I believe, has suspected that these varieties of maize are distinct species; and it is important to notice that the hybrid plants thus raised were themselves perfectly fertile; so that even Gärtner did not venture to consider the two varieties as specifically distinct.

(pp. 269–70)

Darwin then went on to list examples of less than full fertility drawn from maize (or corn), gourds, Verbascum (figworts, often used in herbal smoking mixes), hollyhocks, and tobacco.

One point of considerable interest is that Darwin was adamant that sterility in these cases was not caused by selection. It was always accidental on long separation and selection for other features quite arbitrarily, as side effects, destroying the effectiveness of the reproductive apparatus. Why did Darwin eschew the chance to see sterility in hybrids as the inevitable consequence of selection? The reason he gave was that the causes of sterility seem so varied and so variable – sometimes completely effective and sometimes very ineffective – that no general rules can be drawn. But there is a deeper reason, namely that here selection simply could not be effective, because it would never be of value to something to be sterile if there were no counterbalancing factors (as there are in the case of the social instincts).

The view generally entertained by naturalists is that species, when intercrossed, have been specially endowed with the quality of sterility, in order to prevent the confusion of all organic forms. This view certainly seems at first probable, for species within the same country could hardly have kept distinct had they been capable of crossing freely. The importance of the fact that hybrids are very generally sterile, has, I think, been much underrated by some late writers. On the theory of natural selection the case is especially important, inasmuch as the sterility of hybrids could not possibly be of any advantage to them, and therefore could not have been acquired by the continued preservation of successive profitable degrees of sterility. I hope, however, to be able to show that sterility is not a specially acquired or endowed quality, but is incidental on other acquired differences.

(p. 245)

We have here a reversal of the usual kind of argument. We would not expect natural selection to have done these things, and the evidence is that it did not.

Paleontology

The two chapters on paleontology are not written in the order that I would have written them. Like the song says, I believe in accentuating the positive. I would have made the case – the very strong case – for the explanation of the fossil record by means of evolution through natural selection and then I would have discussed the problems. Almost certainly because the critics were so loud, Darwin did it the other way around, so that by the time you get to the positive you are a little sand-bagged by the difficulties. Faithful to his presentation, therefore, we start with "On the imperfection of the fossil record," where Darwin went over some of the big factors that non-evolutionists had invoked in making their case against change. Of course the biggest problem of all is the ever present issue of gaps in the record. If organisms evolved smoothly from one form to another, where are the intermediates? Why, in climbing up the geological column, do we so often go from one entire form to another, quite different, one? "The explanation lies, as I believe, in the extreme imperfection of the geological record" (p. 280).

In detail, Darwin went over the problems of fossilization, and of organisms ever being laid down, let alone laid down in order and then allowed to remain through the ages, although he did admit candidly to a bit of a circularity about all of this. "I do not pretend that I should ever have suspected how poor a record of the mutations of life, the best preserved geological record presented, had not the difficulty of our not discovering innumerable transitional links between the species which appeared at the commencement and close of each formation, pressed so hardly on my theory" (p. 302). There were some major issues which called for special attention. "The abrupt manner in which whole groups of species suddenly appear in certain formations, has been urged by several palaeontologists, for instance, by Agassiz, Pictet, and by none more forcibly than by Professor Sedgwick, as a fatal objection to the belief in the transmutation of species" (p. 302). The fact is, however, that a group may evolve in one part of the world and then spread around the globe. Although this spreading takes much time in human terms, in paleontological terms it is but an instant and hence would not be inscribed in the record:

> it might require a long succession of ages to adapt an organism to some new and peculiar line of life, for instance to fly through the air; but that when this had been effected, and a few species had thus acquired a great advantage over other organisms, a comparatively short time would be necessary to produce many divergent forms, which would be able to spread rapidly and widely throughout the world.
>
> (p. 303)

Darwin conceded openly the real problem that, as you go down the record, life suddenly appears full-blown; beneath that there is nothing. Where is the evidence of evolution from simpler forms? Perhaps, argued Darwin ingeniously, life first evolved in places now covered by oceans. Remember, Darwin was not a proto-plate tectonicist, thinking that the continents move around the globe altering the positions of the oceans. As a Lyellian, he thought that the world is like (to repeat the modern metaphor) a water bed with some parts rising and other parts sinking and thus altering distributions. Perhaps also, argued Darwin even more ingeniously, the pressure of the oceans on the underlying rocks is so great that everything is now so squashed down that you will never find evidence of life at the earliest stages of earth's history.

> At a period immeasurably antecedent to the silurian epoch [we would now say "Cambrian"], continents may have existed where oceans are now spread out; and clear and open oceans may have existed where our continents now stand. Nor should we be justified in assuming that if, for instance, the bed of the Pacific Ocean were now converted into a continent, we should there find formations older than the silurian strata, supposing such to have been formerly deposited; for it might well happen that strata which had subsided some miles nearer to the centre of the earth, and which had been pressed on by an enormous weight of superincumbent water, might have undergone far more metamorphic action than strata which have always remained nearer to the surface.
>
> (p. 309)

Turn now to the case for evolution. Most significant are the fossil links between two different living forms. You find that in the past there were intermediate forms, suggesting that these then evolved into today's separate groups. Moreover, the greater the time in the past, the more diverse the contemporaneous groups that are linked.

> It is a common belief that the more ancient a form is, by so much the more it tends to connect by some of its characters groups now widely separated from each other. This remark no doubt must be restricted to those groups which have undergone much change in the course of geological ages; and it would be difficult to prove the truth of the proposition, for every now and then even a living animal, as the Lepidosiren, is discovered having affinities directed towards very distinct groups. Yet if we compare the older Reptiles and Batrachians, the older Fish, the older Cephalopods, and the eocene Mammals, with the more recent members of the same classes, we must admit that there is some truth in the remark.
>
> (p. 303)

Even more significant, if evolution be true, is that you expect what is found, namely a roughly progressive record, with the more primitive forms in the older strata and the more advanced forms in the newer strata (Figure 2.4). Natural selection may be relativistic, but not to that extent. "The inhabitants of each successive period in the world's history have beaten their predecessors in the race for life, and are, in so far, higher in the scale of nature; and this may account for that vague yet ill-defined sentiment, felt by many palaeontologists, that organisation on the whole has progressed" (p. 345). Again we have Darwin playing off his own ideas against those of the *Naturphilosophen*. He felt a smug satisfaction in the way that his theory could explain precisely those parallels of which people like Agassiz made so much, especially that between the course of individual life (embryology) and the course of group life (paleontology).

> Agassiz insists that ancient animals resemble to a certain extent the embryos of recent animals of the same classes; or that the geological succession of extinct forms is in some degree parallel to the embryological development of recent forms. I must follow Pictet and Huxley in thinking that the truth of this doctrine is very far from proved. Yet I fully expect to see it hereafter confirmed, at least in regard to subordinate groups, which have branched off from each other within comparatively recent times. For this doctrine of Agassiz accords well with the theory of natural selection. In a future chapter I shall attempt to show that the adult differs from its embryo, owing to variations supervening at a not early age, and being

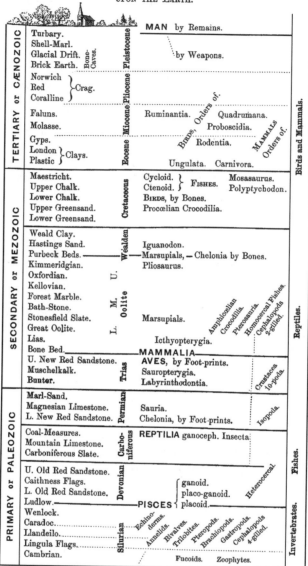

Figure 2.4 The fossil record as known at the time of Darwin's *On the Origin of Species* (Richard Owen, *Palaeontology, or, A Systematic Summary of Extinct Animals and their Geological Relations*. Edinburgh: Adam and Charles Black, 1861, p. 5)

inherited at a corresponding age. This process, whilst it leaves the embryo almost unaltered, continually adds, in the course of successive generations, more and more difference to the adult.

(p. 338)

Thus the embryo comes to be left as a sort of picture, preserved by nature, of the ancient and less modified condition of each animal.

Biogeography

True to expectations, biogeography was a happy topic. The only way in which you can make sense of the distributions of animals and plants is through a naturalistic hypothesis – otherwise God is reduced to exhibiting totally irrational behavior – and that naturalistic hypothesis is evolution through natural selection. There are three important facts that cry out for explanation. First, you simply cannot tie down organisms and their natures to the conditions in which they occur. If you compare North America with South America, for every climate in the one (every ecological niche, as we would say) there is a matching climate in the other. Yet the organisms in the North are distinctively different from the organisms in the South. Second, barriers cause differences. The organisms of Australia, South Africa, and South America, all in the southern hemisphere, are nevertheless as different as it is possible to imagine. Third, where there are no barriers, even though the conditions may be very different, the organisms resemble each other.

Facts like these are clearly accounted for by an evolutionary interpretation. Organisms travel out from where they are, and inasmuch as they are able to travel, they will diversify and change as the conditions dictate, and there will be similarities between adjacent groups. But where there is no possibility of travel, then we have organisms evolving separately and we expect no similarities. There is no reason whatsoever to think that organisms were created more than once in different parts of the world. It is all a question of where they start, where they go to, and where they cannot go to.

Thus far, Darwin approached the topic of geographical distribution in just the way that (above) I regretted he had not approached the topic of the fossil record. Accentuate the positive. But now what of the problems? Most obviously, there is much to be discussed about how species do end up in places where they might not be expected, especially when they seem to have hopped over barriers. Darwin dealt with this issue in much detail, drawing on observations and simple experiments by himself and others, showing how animals and (particularly) plants can be taken from one place to another, despite obstacles in their way. For instance, a lot of seeds can float in salt water and germinate despite long periods of immersion. Birds eat seeds and then often either throw them up or pass them through; in the gap between ingestion and expulsion, the birds could travel long distances. Or seeds and small animals like snails could adhere to the feet of birds and be transported thus. There really are many ways in which organisms can move around the globe.

Detailed discussion was also offered of how the last great Ice Age had affected biogeographical distributions. Already Darwin was starting to move from excuses to explanations, and then, as he came to the end, switched right back into positive mode. It is no surprise, as the following passage shows, that the distributions of organisms on oceanic islands had pride of place.

> Although in oceanic islands the number of kinds of inhabitants is scanty, the proportion of endemic species (i.e. those found nowhere else in the world) is often extremely large. If we compare, for instance, the number of the endemic land-shells in Madeira, or of the endemic birds in the Galapagos Archipelago, with the number found on any continent, and then compare the area of the islands with that of the continent, we shall see that this is true. This fact might have been expected on my theory for, as already explained, species occasionally arriving after long intervals in a new and isolated district, and having to compete with new associates, will be eminently liable to modification, and will often produce groups of modified descendants.
>
> (p. 390)

The kinds of animals we find on islands are those that we might predict: few mammals for instance, because these animals obviously will find it difficult to travel across large tracts of ocean. Interestingly, though, sometimes plants on oceanic islands have adaptations that seem designed to work with mammals – for instance, hooks on their seeds to catch onto the coats of shaggy quadrupeds. None of this, however, quite measures up to the facts of affinity. Why is it that organisms on islands off a continent resemble the inhabitants of that continent rather than another far distant? Darwin naturally spoke of the Galapagos archipelago, on the equator in the Pacific, about 500 miles from South America.

> Here almost every product of the land and water bears the unmistakeable stamp of the American continent. There are twenty-six land birds, and twenty-five of those are ranked by Mr Gould as distinct species, supposed to have been created here; yet the close affinity of most of these birds to American species in every character, in their habits, gestures, and tones of voice, was manifest. So it is with the other animals, and with nearly all the plants, as shown by Dr. Hooker in his admirable memoir on the Flora of this archipelago. The naturalist, looking at the inhabitants of these volcanic islands in the Pacific, distant several hundred miles from the continent, yet feels that he is standing on American land. Why should this be so? why should the species which are supposed to have been created in the Galapagos Archipelago, and nowhere else, bear so plain a stamp of affinity to those created in America? There is nothing in the conditions of life, in the geological nature of the islands, in their height or climate, or in the proportions in which the several classes are associated together, which resembles closely the conditions of the South American coast: in fact there is a considerable dissimilarity in all these respects. On the other hand, there is a considerable degree of resemblance in the volcanic nature of the soil, in climate, height, and size of the islands, between the Galapagos and Cape de Verde Archipelagos: but what an entire and absolute difference in their inhabitants! The inhabitants of the Cape de Verde Islands are related to those of Africa, like those of the Galapagos to America.
>
> (pp. 397–8)

Obviously there is only one sensible way to explain this phenomenon. The organisms of the islands came from their respective neighboring continents, and then evolved into their different and diverse forms.

Systematics, morphology, and embryology

Drawing towards the final chapters of the *Origin*, Darwin moved to a cluster of new topics. First there was systematics or classification. Everyone at the time knew that some classifications seemed more objective, more natural, than others. Grouping humans with the apes is more natural than grouping them with warthogs, for instance. Simply judging on size, grouping whales with elephants, is less natural than grouping cud chewers, cows and giraffes, together. Why is this? Most obviously because of propinquity of descent. All of the issues in classification

> are explained, if I do not greatly deceive myself, on the view that the natural system is founded on descent with modification; that the characters which naturalists consider as showing true affinity between any two or more species, are those which have been inherited from a common parent, and, in so far, all true classification is genealogical; that community of descent is the hidden bond which naturalists have been unconsciously seeking, and not some unknown plan of creation, or the enunciation of general propositions, and the mere putting together and separating objects more or less alike.
>
> (p. 420)

As one who had spent a lot of time studying barnacles, Darwin knew much about the problems of classification. Notably, the features that strike one as the most obvious, those most needed for day-to-day living, are not necessarily the best clues to a natural system. Indeed, they are often the very worst. Darwin's theory had a straight answer for this. Natural selection often grabs a feature and in one branch it does one thing with it and in another branch it does another thing. The surface adaptations do not reveal the underlying links. What you need are those features that go back in time to show original forms, those features that somehow have been constant through time as shown by their constancy today. Out comes the breeding analogy.

> In tumbler pigeons, though some sub-varieties differ from the others in the important character of having a longer beak, yet all are kept together from having the common habit of tumbling; but the short-faced breed has nearly or quite lost this habit; nevertheless, without any reasoning or thinking on the subject, these tumblers are kept in the same group, because allied in blood and alike in some other respects.
>
> (pp. 423–4)

Likewise in nature. Watch out for those analogies that are simply a function of adaptation to an environment as opposed to affinities that reveal descent.

> We can understand, on these views, the very important distinction between real affinities and analogical or adaptive resemblances. . . . The resemblance,

in the shape of the body and in the fin-like anterior limbs, between the dugong, which is a pachydermatous animal, and the whale, and between both these mammals and fishes, is analogical. Amongst insects there are innumerable instances: thus Linnaeus, misled by external appearances, actually classed an homopterous insect as a moth. We see something of the same kind even in our domestic varieties, as in the thickened stems of the common and swedish turnip. The resemblance of the greyhound and racehorse is hardly more fanciful than the analogies which have been drawn by some authors between very distinct animals. On my view of characters being of real importance for classification, only in so far as they reveal descent, we can clearly understand why analogical or adaptive character, although of the utmost importance to the welfare of the being, are almost valueless to the systematist. For animals, belonging to two most distinct lines of descent, may readily become adapted to similar conditions, and thus assume a close external resemblance; but such resemblances will not reveal will rather tend to conceal their blood-relationship to their proper lines of descent.

(p. 427)

None of this meant that Darwin was now giving one inch to the *Naturphilosophen*. The similarities, he said, reveal deep evolutionary relationships, but they are not primary in themselves. Ultimately, they are the product of natural selection, however irritating and misleading the immediate effects of selection may be. I suspect, incidentally, that here we have a major reason why many biologists after Darwin, Thomas Henry Huxley, for instance, were not enthused by natural selection. Much of their time was spent classifying organisms, usually dead, often fossilized. Their daily labors chiefly involved trying to discern the effects of adaptation to new circumstances and then ignoring them!

Morphology was easy for an evolutionist. How does one explain the similarities between the parts of animals (and plants) that are in many respects so different? Similarities of structure that might have no connection whatsoever at the functional level? "What can be more curious than that the hand of a man, formed for grasping, that of a mole for digging, the leg of the horse, the paddle of the porpoise, and the wing of the bat, should all be constructed on the same pattern, and should include the same bones, in the same relative positions?" It is, as Darwin had pointed out earlier, all a matter of Unity of Type. The similarities (what we today, following the anatomist Richard Owen, would call "homologies") reveal shared ancestry (Figure 2.5).

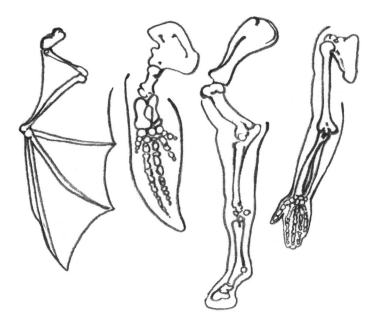

Figure 2.5 Vertebrate homologies: (*left to right*) bat's wing, porpoise's paddle, horse's leg, and human's arm

The explanation is manifest on the theory of the natural selection of successive slight modifications, each modification being profitable in some way to the modified form, but often affecting by correlation of growth other parts of the organisation. In changes of this nature, there will be little or no tendency to modify the original pattern, or to transpose parts. The bones of a limb might be shortened and widened to any extent, and become gradually enveloped in thick membrane, so as to serve as a fin; or a webbed foot might have all its bones, or certain bones, lengthened to any extent, and the membrane connecting them increased to any extent, so as to serve as a wing: yet in all this great amount of modification there will be no tendency to alter the framework of bones or the relative connexion of the several parts. If we suppose that the ancient progenitor, the archetype as it may be called, of all mammals, had its limbs constructed on the existing general pattern, for whatever purpose they served, we can at once perceive the plain signification of the homologous construction of the limbs throughout the whole class.

(p. 435)

On the Origin of Species

Now came embryology, an area that Darwin confessed gave him more pleasure than any other. By the time of the *Origin*, the basic facts were well known. Many species, very different as adults, are frequently very similar as embryos. This is a key fact in classification, for often you cannot work out connections at the adult level, but at the embryonic level all is revealed. More than this: as Darwin had noted in an earlier chapter, in looking at the fossil record, often you find that connections can be made between past and present because the older fossil forms often look more embryonic. Although Darwin mentioned Louis Agassiz, the person who had done most, earlier in the century, to illuminate the topic was the Estonian nobleman (of German descent) Karl Ernst von Baer. Although he was no evolutionist and never became an enthusiast for Darwinism, Baer argued that the embryos of earlier forms are like the embryos of later forms; further, that often the earlier forms do not change much in individual development, so the adults of earlier forms resemble the embryos of later forms, even when the later forms change a great deal in individual development. Although his thinking in some respects overlapped with that of the *Naturphilosophen*, in other respects Baer was far from their way of thinking. He saw no inevitable upward progress, but a branching development from one original embryo type in different ways to different adult forms.

Despite Baer's anti-evolutionary thinking, one can easily see how Darwin drew confidence for his own thinking from such a picture of development. Indeed, fleshing out what he had said earlier, Darwin's solution to the mysteries of embryology was simple. Early forms do not show much change during development. Then evolution starts to occur. New features, new variations, often come at times when the individual organism is well along the way of its development. If they matter, natural selection picks up on these. In the embryo, as such, there is little need for change, and there is little selective competition, so the earliest changes will generally not get picked up. But later, out of the womb, selection will start to count and to separate organisms, as different new variations do get picked up. Again the breeders' world was important here, for no one much cares about the juveniles. It is the adults which count.

Some authors who have written on Dogs, maintain that the greyhound and bulldog, though appearing so different, are really varieties most closely allied, and have probably descended from the same wild stock; hence I was curious to see how far their puppies differed from each other: I was told by breeders that they differed just as much as their parents, and this, judging by the eye, seemed almost to be the case; but on actually measuring the old dogs and their six-days old puppies, I found that the puppies had not nearly acquired their full amount of proportional difference. So, again, I was told that the foals of cart and race-horses differed as much as the full-grown animals; and this surprised me greatly, as I think it probable that the difference between these two breeds has been wholly caused by selection under domestication; but having had careful measurements made of the dam and of a three-days old colt of a race and heavy cart-horse, I find that the colts have by no means acquired their full amount of proportional difference.

(pp. 444–5)

The reason is obvious: "Fanciers select their horses, dogs, and pigeons, for breeding, when they are nearly grown up: they are indifferent whether the desired qualities and structures have been acquired earlier or later in life, if the full-grown animal possesses them" (p. 446). Exactly the same sort of thing happens in nature. "On this view we can understand how it is that, in the eyes of most naturalists, the structure of the embryo is even more important for classification than that of the adult. For the embryo is the animal in its less modified state; and in so far it reveals the structure of its progenitor" (p. 449). Moreover: "As the embryonic state of each species and group of species partially shows us the structure of their less modified ancient progenitors, we can clearly see why ancient and extinct forms of life should resemble the embryos of their descendants, – our existing species" (p. 449). Note that this is not simply a forerunner of Haeckel's biogenetic law. Haeckel's law has the flavor of *Naturphilosophie* and Darwin wanted nothing of the inevitable, straight-line momentum of that kind of thinking.

Conclusion

There followed a quick discussion of rudimentary and atrophied organs, good clues to what happened in the past, and the work of the *Origin* was over. A summary, the quick reference to our own species, and Darwin was ready for his final flowery paragraph.

It is interesting to contemplate an entangled bank, clothed with many plants of many kinds, with birds singing on the bushes, with various insects flitting about, and with worms crawling through the damp earth, and to reflect that these elaborately constructed forms, so different from each other, and dependent on each other in so complex a manner, have all been produced by laws acting around us. These laws, taken in the largest sense, being Growth with Reproduction; inheritance which is almost implied by reproduction; Variability from the indirect and direct action of the external conditions of life, and from use and disuse; a Ratio of Increase so high as to lead to a Struggle for Life, and as a consequence to Natural Selection, entailing Divergence of Character and the Extinction of less-improved forms. Thus, from the war of nature, from famine and death, the most exalted object which we are capable of conceiving, namely, the production of the higher animals, directly follows. There is grandeur in this view of life, with its several powers, having been originally breathed into a few forms or into one; and that, whilst this planet has gone cycling on according to the fixed law of gravity, from so simple a beginning endless forms most beautiful and most wonderful have been, and are being, evolved.

(pp. 489–90)

3

One Long Argument

In England, the reception of the *Origin* by the philosophers of Darwin's day ranged from outright hostility to lukewarm acceptance mixed with misunderstanding (Ruse 1979). William Whewell, by now Master (principal) of Trinity College, Cambridge, supposedly refused to allow it to appear on the shelves of the library. He added a fiery little note of dissent to the next edition of a work on natural theology that he had written thirty years before, his *Bridgewater Treatise* on astronomy. John Herschel referred to the *Origin* as the theory of "higgledy-piggledy." He did speculate privately about a saltationary theory of evolution, but it was not a Darwinian theory. John Stuart Mill, who was by now Britain's leading philosopher, seemed more warmly inclined to the *Origin*. But in fact his praise was very limited, refusing to allow that Darwin had really come up with a fully established position.

> Mr. Darwin's remarkable speculation on the *Origin of Species* is another unimpeachable example of a legitimate hypothesis. What he terms "natural selection" is not only a *vera causa*, but one proved to be capable of producing effects of the same kind with those the hypothesis ascribes to it: the question of possibility is entirely one of degree. It is unreasonable to accuse Mr. Darwin (as has been done) of violating the rules of Induction. The rules of Induction are concerned with the conditions of Proof. Mr. Darwin has never pretended that his doctrine was proved. He was not bound by the rules of Induction, but by those of Hypothesis.
>
> (Mill 1874, 328)

But of course, Mr Darwin was very much into the proof business, so Mill was offering half a cake at the most.

Darwin in his autobiography referred to his work as "one long argument," and this is the right way to think of it. The *Origin* is not simply one idea after another, strung together in haphazard fashion between two covers. Although, in the end, Darwin wrote rapidly, he was drawing on thinking that went back twenty years. Indeed, the structure of the *Origin* is really not that different from the earliest sketch of 1842. What is a tad misleading is the almost casual, friendly style of the *Origin*. One does not think of this as a great work of science – more a popularization or something like it. In part, this comes from Darwin's inability to put anything in mathematical form, the sure mark today of professional science. In part, it comes from his background. It had been his father and his uncle, Emma's father, who had sponsored and supported him. As with the book on the *Beagle* voyage, it was natural for Darwin to write for such an audience. The ways he learnt and adopted as a young man stayed with him always. We must be careful not to underestimate the beast beneath the smooth, surface form.

Structure

Like Gaul, the *Origin of Species* is divided into three parts. First, there is artificial selection and the analogy to natural selection. Second, there is natural selection itself and its derivation via the struggle for existence from the Malthusian ratios. Third, there is the application of evolution through natural selection to the many areas of biology – instinct, paleontology, biogeography, and so forth. There is more to the *Origin* than this. There is the question of natural variation and how it is needed to get natural selection. There is the question of heredity and how variations must persist through the generations if they are to have lasting evolutionary effects. There is the repeated, direct use of artificial selection to illustrate points about the subdivisions of biology, as we have seen it used in the embryological discussion. And there are the connections between the subdivisions themselves: for instance, the way that embryology is used to explain some of the facts of paleontology. So, I am not pretending that the threefold division is complete, but it will give us enough to go on for now.

To start with the most basic question: what kind of theory is Darwin offering us? We need not work in the dark here, for much has been written by philosophers and others about the nature of scientific theories and we can draw on these discussions. The classic picture is that of a theory as a "hypothetico-deductive system," or "H-D system", for short (Nagel 1961; Hempel 1966). Here the ideal is Newtonian mechanics. You start with a number of basic laws or hypotheses. They are not logically necessary or even *a priori* (meaning acquired independently of sense experience). In fact, they are anything but this, because science is about the real world. But they are basic within the system: they are the premises. Often it is thought that these hypotheses make reference to unseen entities, like genes or molecules, although I am not sure that this is an essential feature of an H-D system. What is important is that everything else is shown to follow, deductively, from the premises. And the lower-level derivations (often called the "empirical laws") are about real, observable entities. Thus, in Newton's system, we start with his three laws of motion and his law of gravitational attraction, and then we can derive Galileo's laws about the nature of falling bodies (as, projected missiles describe parabolas) as well as Kepler's laws of planetary motion (as, all planets travel in ellipses with the sun at one focus).

I should say that the H-D picture has been much criticized in the past forty years (Giere 1988). It is often referred to contemptuously as the "received view," meaning the "not-received-by-anybody-of-good-sense-including-me view." Critics often want to substitute what they call the "semantic view," where science is seen less as building grand theories and more as taking bits of theory, running up idealistic models, and then seeing if they apply in the real world. So if you have a problem with cannons and their firing, you turn to Galileo's work and try to build a model that takes wind resistance into account, and then see if that fits your empirical findings. For myself, I am inclined to take a somewhat pragmatic view of these issues. If you are talking about what scientists actually do, then building models and seeing if they apply does seem to characterize the day-to-day activity. When Darwin came up with his coral reef theory, he had a problem and proposed a little model within the Lyellian system – sinking islands and building corals – and then supposed that it applied to the cases in hand. However, as an overall view of what a theory looks like – often, in fact, a highly idealized view of what a theory looks like – then the H-D picture does tie things together and make good sense. More than this, it is worth asking what kind of theory Darwin thought that he had. It is true that one does not necessarily produce what one aims to produce, or even what one thinks one has produced, but it is a good start to the questioning.

Judged in these terms, there is overwhelming evidence that Darwin held an H-D picture of theories, if not by this name. Newton was the ideal of theorist and we have seen that this ideal was accepted by Darwin. There is no surprise here, for Darwin was sensitive to scientific methodology – so much so that there was good reason why he was hurt by the reactions of the philosophers. Darwin had read, liked and, by his own admission, been much influenced by the *Preliminary Discourse on the Study of Natural Philosophy* by the philosopher and astronomer Herschel. The work was published in 1830; Darwin read it just after graduating in 1831. (Note that, in those days, "natural philosophy" meant "natural science." "Moral philosophy" was what we mean by "philosophy.") Darwin also talked much to Whewell (who recommended and reviewed Herschel's book), and after the *Beagle* voyage Darwin read (twice) Whewell's *History of the Inductive Sciences*, as well as a very long review of Whewell's *Philosophy of the Inductive Sciences* written by Herschel. After the voyage, Darwin mixed with Whewell and others interested in methodology, like Lyell and the inventor of the computer Charles Babbage. In other words, he was soaking up information about scientific methodology just when he was thinking hard about the problem of evolution.

All of these people were deeply committed to the hypothetico-deductive system, and all of the evidence suggests that Darwin accepted it also. Herschel (1830) was explicit.

> The whole of natural philosophy consists entirely of a series of inductive generalizations . . . carried up to universal laws, or axioms, which comprehend in their statements every subordinate degree of generality, and of a corresponding series of inverted reasoning from generals to particulars, by which these axioms are traced back into their remotest consequences, and all particular propositions deduced from them.
>
> (p. 104)

Likewise, here is Whewell (1837) on Newtonian mechanics: "Newton's theory is the circle of generalization which includes all the others; the highest point of the inductive ascent; the catastrophe of the philosophic drama to which Plato had prologized; the point to which men's minds had been journeying for two thousand years" (2, 183).

Darwin rather cheekily quoted Whewell (1833) at the beginning of the *Origin*: "But with regard to the material world, we can at least go so far as this – we can see that events are brought about not by insulated interpositions of Divine power, exerted in each particular case, but by the establishment of general laws."

The argument for natural selection

So what parts of the theory of the *Origin* are H-D? Well, obviously nothing very much, if you are going to insist on a strict definition. But if you think about the kind of argument Darwin puts forward in the book, good candidates are those passages dealing first with the struggle for existence and then moving on to natural selection. Darwin begins with the Malthusian ratios. "Population numbers go up potentially geometrically." "Food and space supplies go up potentially arithmetically." "Geometric progression is greater than arithmetic progression." (This one is assumed and not stated.) Hence: "There will be ongoing struggles for existence." (If you like, add some qualifiers about the impossibility of "prudential restraint.") Then: "All natural populations show variation." (Note, this is assumed to be heritable variation and that more variations will always come.) Hence: "Some will get through and some will not, and there will be differences between the winners and losers and these differences will, on average, be crucial to success or failure." (Some hidden assumptions here: wait on these for a moment.) Call this natural selection. (Definition.) Repeat the process many times and the result is full scale evolution. (A kind of integration or summation.)

There is nothing very formal here, obviously. But if you are fairly charitable, you can say that there is the sketch of some kind of axiomatic argument at work. Of course, where you go after that is another matter. Even leaving the analogy from artificial selection to one side, the H-D picture seems less and less appropriate – or perhaps one might say more and more sketchy. Taking this second, softer, line, one avoids saying that the H-D picture is absolutely wrong (which would certainly have upset Darwin). One says merely that the discussion of the *Origin* gets more and more loose or casual. Suppose you are thinking about the Galapagos finches. It is certainly not fair to say that Darwin simply assumes that they are going to evolve by natural selection. There is a lot of discussion, some explicit some more implicit, about the ways in which the finches' ancestors might have been blown across the sea from South America to the Galapagos archipelago, and how then they might (or might not) have hopped from island to island. And about the different conditions that they faced on the islands from the mainland. Also about the ways in which sterility barriers are thrown up. And so on and so forth. But you are certainly not presented with anything like a formal deduction, even though it is clear that Darwin thinks that all of the discussions in the last part of the book are, in fact, related through – and explained by – premises about selection. A sketch, although no more than a sketch.

There is more to discuss about these central arguments. They are supposed to be made up of laws. Is this indeed the case? A lot of people worry that it is not so (Elgin 2006). They point out that the trouble with biology is that every time you try to think up a law, something goes wrong. It is not just Darwin's problem: it is a general problem. You can never have laws in biology. This, to put it mildly, is a pretty strong charge and needs a reply. So ask first: what is a law? Most people would accept something like this: "Laws of nature are true, universal statements about the world." Hence, "Charles Darwin was an Englishman," is no law because it is singular, and "All Englishmen are Jewish" is no law because it is false. And "Everything is identical to itself" is no law of nature because it is a logical truth, and holds irrespective of the nature of the physical world.

However, there is more than this. "All people called Michael Ruse are philosophers" is to the best of my knowledge true. (To our great mutual surprise there are two of us with the same name. We are not related.) But even if the Ruse–philosopher connection is true, and even though it is both universal and about the empirical world, it is not a law. Somehow, it is not necessary. One of my sons might have a son and (unlikely as it now seems to any of us) call him after me. That new Michael Ruse might think that philosophy is a silly subject. Why not? Everyone else in my Ruse family feels this way. The point is that laws are necessary. (Sometimes this is called "nomic" necessity to distinguish it from logical or mathematical necessity. There is no great magic here. It is just a name, taken from the Latin for "law.") If we say that "Planets go round the sun in ellipses," we mean that in some way they have to. A planet simply cannot describe a figure of eight. Often the necessity of laws is expressed as an ability to bear counterfactuals. We might say: "That celestial object cannot be a planet because it is not describing an ellipse." To the contrary, someone called Michael Ruse might not be a philosopher.

Some people argue that laws must be true – no exceptions. As with the H-D system, I feel somewhat more relaxed about this. There is something a bit odd about saying that something is both necessary and not always true. So say, rather, that I am prepared to build in a fairly significant *ceteris paribus* clause into laws. All other things being equal, laws must be true, but often in the real world all other things are not equal. If I see an object refuse to fall to the ground – a plane or a boomerang – I look for reasons why not. It must fall to the ground unless something is preventing it. Objects fall to the ground, all other things being equal, but when the air is pushing them up all other things are not equal. Likewise, suppose I am listening to "Car Talk" on National Public Radio, and one of the Magliozzi brothers, Ray or Tom, tells us that it could not be a double-ended Boston sprocket that is causing the caller's car to leak oil, because double-ended Boston sprockets were not introduced into the USA until 1961, and she is driving a '58 Chevy. I am not going to turn off the set or lose faith in the order of nature if it turns out that her Chevy was a special order and a one of its kind with the new sprocket. I guess, however, that, much faith though I have in the infallibility of the Magliozzi brothers, I am going to be rather less surprised if their claims about cars prove wrong than if claims about gravity prove wrong. Nomic necessity is not something we find out there in nature. It is more our way of conceptualizing the world. These things come in degrees and we know that some things are less exception prone than others.

It is the same with Darwin's general claims. Biology tends to be a bit "softer" than physics. There are more exceptions in biology than there are in physics, even though there are more exceptions in car design than there are in biology. But this said, it does seem to me that Darwin's claims fall on the side of law. Overall, he based his thinking on generalizations about empirical reality that are true and not just contingent – they are, in some sense, necessary. And this is more or less so, even if there are some real or possible counter-examples. Organisms do have a tendency to go up in numbers geometrically, and food and space do not. There are struggles for existence. As Darwin himself admits, not always or at all times, but generally so. "Although some species may be now increasing, more or less rapidly, in numbers, all cannot do so, for the world would not hold them" (Darwin 1859, 64). Likewise with variation. Populations do contain lots of variation. That is a fact. You could produce an absolutely standard group – when it comes to testing organisms like mice, drug companies and others often have need of just that – but what you find in nature is variation. And none of this is mere chance. If I were a professor of biology, and some student came back after a summer field trip and told me that the gophers in his study showed no variation, I would think he was simply not doing the job properly – or that my friends were playing a joke on us.

Natural selection as a scientific concept

But what about natural selection itself? Is it a genuine scientific concept or notion? A lot of people worry about it (Sober 1984). It is such a simple notion – Huxley kicked himself for not having thought of it on his own – and yet it seems to do so much. Surely there is something phony here? A conjurer's trick or sleight of hand? Start with the question about whether or not it is a force, as Darwin thought. Can this be so? In a way, you might well think not. Natural selection does not exist – at least, not like a planet or a molecule. You have (say) a population of organisms, half white and half black. A predator eats more black ones than white ones because the black ones stand out, and in a few generations the population is nearly all white. You say that a selective "force" was at work, but really all you are doing is abbreviating the statement that the black stood out and was eaten and the white did not. There was nothing over and above this. In a world where we did not need shortcut devices and where Occam's razor ruled okay, we would not talk of selection at all.

Well, in response, we do live in a world where we need shortcut devices and Occam's razor sometimes shaves too close. Hurricane Katrina did terrible damage to New Orleans. You could say that it was all a matter of wind and rain – molecules moving at a certain speed and in a certain direction, if you really want – but ordinary people can still say that the city was hit by a terrible force of nature. It is a bit like saying that I am sitting at a table, as opposed to sitting at a buzzing, booming configuration of molecules, as the physicist Arthur Eddington (1929) was wont to say, and insisting that the table is just as real as the molecules, as Arthur Eddington was not wont to say. More than this, of course: if you identify the configuration as a force, as a hurricane, then you can start to compare and contrast with other such phenomena. Was Katrina really bad or was it a case of inadequate protection, and so forth? Similarly with natural selection. Do we get forces of similar type with populations of red and green organisms, for example? No one could claim that Darwin's force is identical to Newton's force – there is no law of inverse, squared distance, for example – but I do not see why it should not be called a "force." Whether it is unobservable in principle seems to me a judgment call. Obviously, selection was unobservable by us when it operated back in the Cambrian era, and is unobservable by us now at the micro-level. Perhaps you want to say that we always see the effects of selection. For myself, when I see a bird with a red caterpillar in its beak and a green one still on my young tomato plants, I say that I have just seen natural selection at work.

Notice, however, that natural selection does introduce a special way of thinking into biology – namely, statistical thinking. One is now working with groups and averaging over them. One is not saying that individual things or events have no causes, and one is certainly not saying, as in quantum mechanics, that there is a level beyond which we cannot peer when it comes to causes, but one is ignoring individual causes and thinking in terms of groups. It may be that this black individual was spotted by the predator purely by chance and only after a long search, but these differences are ironed out and ignored: the question is about what proportion of blacks are taken over whites and so forth. Fortunately for Darwin, he did not have to fight a battle on this front, because he was working just at a time when the physicists were also introducing statistical methods into their science. The American pragmatist philosopher Charles Sanders Peirce spotted this.

The Darwinian controversy is, in large part, a question of logic. Mr. Darwin proposed to apply the statistical method to biology. The same thing has been done in a widely different branch of science, the theory of gases. Though unable to say what the movements of any particular molecule of gas would be on a certain hypothesis regarding the constitution of this class of bodies, Clausius and Maxwell were yet able, by the application of the doctrine of probabilities, to predict that in the long run such and such a proportion of the molecules would, under given circumstances, acquire such and such velocities; that there would take place, every second, such and

One Long Argument

such a number of collisions, etc.; and from these propositions they were able to deduce certain properties of gases, especially in regard to their heat-relations. In like manner, Darwin while unable to say what the operation of variation and natural selection in every individual case will be, demonstrates that in the long run they will adapt animals to their circumstances.

(Peirce 1877, 3)

As it happens, Peirce was never a fan of Darwin's theory, thinking in 1893 that "to a sober mind its case looks less hopeful now than it did twenty years ago," and that even back then it "did not appear, at first, at all near to being proved" (Peirce 1893, repr. 1935, 6, 297). But he had no objections to the group approach taken by Darwin.

There is still a worry – some would say the biggest worry – about selection. If natural selection is equivalent to the survival of the fittest, then isn't this just the same as saying that natural selection is tautological? Who are the fittest? Those that survive! Hence, natural selection is really just a fancy redescription of the phenomenon – those that survive are those that survive. It is not an empirical notion at all. A lot of people have argued this, and not all of them enemies of science. For a long time, the Austro-British philosopher of science Karl Popper (1974) used to think that Darwinism was not a genuine theory but a "metaphysical research programme," because natural selection was a tautology. In response, there are certainly some issues here, but nothing like as devastating as the critics claim. How could they be? Populations do have differences and those differences do make for changes in proportions: these are true empirical claims, so selection cannot just be a tautology. If a predator is eating more blacks than whites and the reason is that the whites have better camouflage against a pale background than do the blacks, there is no tautology in saying that this is so. The claim may be false; it is not necessarily true.

Part of the problem here is a confusion about the nature of theories, and given our discussion above we can clear this up quickly. If a scientist is running up a model, then at one point no one is talking facts: the model is theoretical. So if you say that greens are selectively favored over reds, then clearly you are just stipulating conditions: greens must beat out reds in your model. But now there is the empirical job of seeing if your model actually applies in nature. Do we find populations approximating the reds and the greens? If we do, then well and good. And if not, run up another model. All of this is more than just theory, and shows an empirical level to selection studies.

Part of the problem is a bit more subtle, and goes back to our hurricane example. Once we start to think about the hurricane as a something, as a force, we can get into the business of comparing and contrasting, and this is the beginning of science. Without laws, without generalities, you can achieve nothing. (The thinking does not have to be about forces. Tables will do as well.) Once evolutionists identify natural selection at work, then they too can start to compare and contrast. Is the black/white predation situation just like the red/green predation situation, for instance? And if it is not, then evolutionists can and will jump right in and try to find out why. Perhaps the reds get taken because they just stand there dumbly, waiting to be eaten. Perhaps the whites do not get taken in such numbers because they are cunning and good at hiding, or some such thing. What you have operating here is some kind of inductive and causal assumption. You expect the same effects given the same causes. This underlies your thinking about selection and perhaps explains why Popper – who could not abide induction – had so much trouble with the concept. For the rest of us, however, simple though natural selection may be, tautological it is not.

Verae causae

Move on to the next question. If natural selection is a force of some kind, it is a cause. It makes other things (effects) happen. But is it a well-established cause? Perhaps New Orleans was demolished because it was a modern-day Sodom or Gomorrah and had incurred God's wrath. But I suspect that most of us would not think this an adequately established cause. Paris in the 1890s was left standing. What of natural selection? Here it really is helpful to put Darwin into the context of his time. He knew he had a major problem. He had to persuade people that natural selection was a cause, even though most of the time its supposed actions were in the past and hence most of the time they could not see it. How can you persuade someone that natural selection was causing evolution back in the Cambrian? Darwin did not have to work blind at this point, because the methodologists of science – Herschel and Whewell – were laying out the conditions (Ruse 1975). Newton had somewhat mysteriously argued that the best science is based on what he called "true causes," *verae causae*. The question now is what exactly is a true cause? Neither Herschel nor Whewell would have felt comfortable with the usage of John Stuart Mill in the passage at the beginning of this chapter, where he used the term for something he did not think well established. For them, it must be true. But the grounds for thinking it true divided the two men.

Herschel (1830) took more of an empiricist attitude, arguing that a true cause is either something we have sensed directly or something we have analogous evidence of having sensed directly. He argued that we know that there is a force keeping the moon circling the earth because we have all experienced the force on a piece of string when we whirl a stone around our hand. He praised Lyell's theory of climate precisely because we have experience of the Gulf Stream and its warming effects on Britain. Whewell (1840) took more of a rationalist attitude, arguing that a true cause does not have to be sensed or experienced: its existence is proved by what it can do. Can it explain many different things? If it can, then we can assume its existence, whether or not we have any direct knowledge of it. Whewell stated that a true cause emerges from what he called a "Consilience of Inductions."

> Accordingly the cases in which inductions from classes of facts altogether different have thus *jumped together*, belong only to the best established theories which the history of science contains. And as I shall have occasion to refer to this peculiar feature in their evidence, I will take the liberty of describing it by a particular phrase; and will term it the *Consilience of Inductions*.
>
> (Whewell 1840, 2: 230)

The difference between the two notions of *vera causa* can be seen in the thinking of a detective. The body lies in a pool of blood in the study. Who killed Sir Redvers Featherstonehaugh-Cholmondeley Bt (pronounced "Reevers Fanshaw-Chumley," I kid you not)? On Herschellian grounds, the crime is pinned on the butler, either because the gardener was looking in the window and saw the butler plunge in the knife, or because the butler had recently been released from prison where he had served time for stabbing his last employer in a similar fashion. On Whewellian grounds, the crime is pinned on the butler because the knife was owned by the butler; he lied about his whereabouts at the time of the crime; he had a cut on his hand and his blood group matched a stain on the victim's coat; his daughter had been seduced by the baronet and so there was a motive; and so forth. There was a mass of circumstantial evidence.

I suspect it is a bit of a toss up, or a matter of situations, as to which method of confirmation most of us think is the more reliable. Few today would think that eyewitness testimony is always better. When judging a rape case, with the option of getting DNA fingerprints, and given the knowledge that testimony under stress may be unreliable, which would you prefer to have – the fingerprints or the testimony? Back in the time of Darwin, the important thing is that the dispute over the correct meaning of *vera causa* was no academic debate. Geology obviously was involved. Someone who takes the Lyellian actualist position is right in Herschel's camp, insisting on causes that have been and can be experienced today. Someone who takes the catastrophist position is, as Whewell pointed out nonstop, justified inasmuch as he can point to phenomena that can be explained (and probably only explained) by massive upheavals. As pertinently, there was a major debate over the theory of light. Given the brilliant experiments at the beginning of the century, especially by Augustin Fresnel, all now wanted to justify a move from the particle theory of Newton to the wave (or, as it was known, "undulatory") theory of Christiaan Huygens. You do not see light waves directly, and the attempt to explain things like interference patterns led Herschel into mind-contorting thought experiments with tuning forks and sealing wax and all sorts. Whewell, somewhat smugly, eschewed all of this and calmly endorsed waves as a cause, lying at the heart of a consilience of inductions. End of argument.

Darwin, bright young man that he was, absorbed and used both notions of true cause! First, there was the analogy from the domestic world of selective breeding. Here we have a force that we ourselves experience, that we ourselves cause, and it makes natural selection plausible and probable. Then, evolution through natural selection is seen to be at the heart of a consilience, with instinct, paleontology, biogeography, and so forth being explained by selection and in turn making a belief in selection plausible. Darwin knew what he was up to. Again and again, he defended the plausibility of his theory on both *vera causa* grounds. The following letter to the botanist George Bentham (nephew of Jeremy Bentham) is typical.

One Long Argument

In fact the belief in natural selection must at present be grounded entirely on general considerations. (1) on its being a vera causa, from the struggle for existence; & the certain geological fact that species do somehow change (2) from the analogy of change under domestication by man's selection. (3) & chiefly from this view connecting under an intelligible view a host of facts.

(Letter of May 22, 1863; Darwin 1985–, 11, 433)

The question we want answered now is how successful Darwin was in his strategy. Historically speaking, as we know, only partially. Why was this? Well, for a start, the breeding analogy really did not convince. People agreed that breeders could do a lot (although one suspects that those not as well acquainted with agriculture as was Darwin would not realize how much breeders could do), but they did not think that breeders could do the kinds of things that Darwin's theory needed. In fact, paradoxically, until the *Origin*, the evidence of the breeders' world had been used against evolution! Wallace (1858) spent much space in the essay he sent to Darwin explaining why the breeders were no true analogy. Everyone simply said that you cannot turn a cow into a horse. For good measure, Huxley (reviewing the *Origin* in 1859) added the fact that no one had yet achieved reproductive isolation between hitherto interbreeding groups, and until they did he for one was going to suspend judgment on natural selection. He did not himself set out to perform experiments to see if it could be done, but then he was not that interested in seeing natural selection vindicated.

The consilience was clearly more successful. This does seem to have convinced people of evolution. The circumstantial evidence was simply massive. If you were going to accept any kind of naturalistic position – that is, to think that organisms were produced by unbroken law and not by miracles – then Darwin had made the case. It was no longer reasonable to believe otherwise. The Galapagos birds are like the South American mainland birds and not like those of Africa. Why, other than through evolution? Humans, cows, birds, porpoises do have bones in common, even though the functions are different. Why, other than through evolution? Embryos do look alike and ancient forms are more embryonic. Why, other than through evolution? Yet it is clear that the consilience was only partly successful. People agreed that natural selection could do something, but few agreed that it could do as much as Darwin wanted. So long as the birds changed somewhat when they went to the Galapagos, that was enough, however it was caused. So long as the ancestor of humans and cows and the others existed, that was enough, however it was caused. And the same is true of the rest. Evolution, yes, but natural selection, only maybe.

It is not easy to ferret out the exact reason why John Stuart Mill was so cool towards Darwin's work. One suspects that, as an empiricist, he just did not like consilience, at least not judged as a ground for proof. I suspect that many today would think that he was wrong here. Even though (for the kinds of religious reasons to be discussed in a future chapter) Whewell did not accept Darwin's theory, he was right about the strength of a good consilience. We may not want to hang the butler on the strength of the circumstantial evidence, but we are prepared to convict and incarcerate for a very long time. The pertinent historical point is obviously that, although people judged Darwin's consilience as good enough for evolution, they drew back when it came to his putative cause.

Metaphor

Let us end by going back to the point where we came in: Darwin's almost casual style. This was noted by the reviewers, particularly the anatomist and foe of the Darwinians, Richard Owen (1860), who sneered that Darwin was writing less in the mode of a professional scientist than in the mode of a popular travel writer. Many seized on the almost exuberant use of metaphor – struggle for existence, natural selection, division of labor, tree of life, and so on – and to an extent this put Darwin on the defensive. We saw in the last chapter that he realized that he used metaphor, and from the beginning he was edgy about his metaphors. This was a reason for introducing the term "survival of the fittest" as a synonym for "natural selection." In later editions of the *Origin*, Darwin explained and justified himself at length.

> In the literal sense of the word, no doubt, natural selection is a false term; but who ever objected to chemists speaking of the elective affinities of the various elements? – and yet an acid cannot strictly be said to elect the base with which it in preference combines. It has been said that I speak of natural selection as an active power or Deity; but who objects to an author speaking of the attraction of gravity as ruling the movements of the planets? Every one knows what is meant and is implied by such metaphorical

One Long Argument

expressions; and they are almost necessary for brevity. So again it is difficult to avoid personifying the word Nature; but I mean by Nature, only
the aggregate action and product of many natural laws, and by laws the
sequence of events as ascertained by us. With a little familiarity such
superficial objections will be forgotten.

(Darwin 1959, 165; this passage was added
in the third edition, 1861)

Was Darwin at fault in using so many metaphors? It cannot be
denied that there are those, both scientists and philosophers, who feel
uncomfortable about it (Fodor 1996). They look upon metaphor as,
at best, shorthand for that which can be said literally and, at worst, a
lazy substitute for that which should be said literally. For these critics,
perhaps the most satisfactory way of looking at metaphor in science
is that it is a sign of immaturity, and that as the science develops
and grows to adulthood it will drop the metaphors. So, in the case
of natural selection, for instance, one should think of "differential
survival" or (even better) "differential reproduction," which is a non-
metaphorical concept and much more precise anyway. Against this,
however, there are those – and I am one – who think that this is all
a little bit too hasty (Lakoff and Johnson 1980; Ruse 1999). First,
even if one could eliminate metaphor, it would be a silly move to make.
Second, there are theoretical reasons to think that the metaphors could
not go without significantly altering the science itself.

Focusing first on the more pragmatic factors, our argument is that
metaphors have an incredible heuristic power and that without them
science as we know it would grind to a halt. Think, for instance, of
physics, and the metaphor of electricity as a fluid. By thinking of it
like this, all kinds of relationships were uncovered – voltage is akin to
water pressure, a battery is like a pump or a reservoir, the current is
the flow – and then they could be given formal treatment. It is true
that, for the very reason that metaphors are not literal, there is always
the danger of misrepresenting what is happening: electricity is not really
a fluid, and you can drill right through a copper rod and the electricity will not leak out; but danger is the price one pays for advance
and understanding.

In Darwin's case, it is obvious that the tree of life metaphor stimu-
lated him to think about causes for branching. Evolutionist predecessors
like his grandfather never thought of change in quite this way and so
they never speculated on the causes of speciation. Even more powerful

was the metaphor of a division of labor. Darwin used it all of the time, applying it to the individual and to the group, to one organism and to another. It was the clue leading to Darwin's principle of divergence and his theory of speciation. Let us repeat the passage, looking now at the language for its own sake:

> The advantage of diversification of structure in the inhabitants of the same region is, in fact, the same as that of the physiological division of labour in the organs of the same individual body – a subject so well elucidated by Milne Edwards. No physiologist doubts that a stomach adapted to digest vegetable matter alone, or flesh alone, draws most nutriment from these substances. So in the general economy of any land, the more widely and perfectly the animals and plants are diversified for different habits of life, so will a greater number of individuals be capable of there supporting themselves. A set of animals, with their organisation but little diversified, could hardly compete with a set more perfectly diversified in structure.
>
> (Darwin 1859, 115–16)

The metaphor also played a role in the explanation of different castes in the Hymenoptera. Darwin wrote about how:

> the wonderful fact of two distinctly defined castes of sterile workers existing in the same nest, both widely different from each other and from their parents, has originated. We can see how useful their production may have been to a social community of ants, on the same principle that the division of labour is useful to civilised man. Ants, however, work by inherited instincts and by inherited organs or tools, whilst man works by acquired knowledge and manufactured instruments.
>
> (pp. 241–2)

We shall see in a later chapter how this metaphor of a division of labor continues, even to this day, to have a crucial role in work on the Hymenoptera.

This is not to say that metaphors necessarily are always helpful. I stress again that metaphors by their very nature are not literal, and there is always the danger of reading too much into one, thinking that a feature of the metaphor must be a feature of the reality. Take, for instance, the notion of the tree of life. At once one is directed to thinking that there is just one original starting place for life, and, moreover, that as soon as a split occurs then the different branches are forever sundered. These assumptions may be true, but there is no logical reason why they have to be, nor is it necessarily a part of selection theory (including the very fact of evolution) that they should have to be. One can easily think of life forming on different parts of the globe and never really combining. One can also conceive of ways in which hybridization might be very important in the history of life, and hence that the branches could rejoin later, after they have split. I shall be speaking more to some of these issues in later chapters. The point here is not that metaphors always mislead, but that one must be careful to see that they do not mislead.

Function

Let us agree, then, that metaphors have a pragmatic role. What about the second question, about their possible implication for the deeper levels of a theory, for the very structure and logic of a theory? The key metaphor in the whole of the *Origin* is that of design (Ruse 2003). Natural theology (about which much more will be said in a later chapter) argues that the world, especially the living world, is not just thrown together. It is organized, it functions, it is as if designed. This insight goes back to the great Greek philosophers Plato and Aristotle. The latter, who was incidentally a fine practicing biologist, talked of causes that bring things about, "efficient causes," and causes for the sake of which things exist, "final causes." The eyes develop because of one or more efficient causes. The eyes exist because of a final cause, in order to see. The great Christian philosophers Augustine and Aquinas took Greek thought and gave it a place in their theology. Final causes spell design. The telescope exists in order that we should be able to see distant objects. There had to be a telescope designer. Telescopes do not happen by blind chance. Therefore, the eyes had to have an eye designer. Things like eyes do not come about by chance. Hence, there is reason to believe in the Great Optician in the Sky.

Leaving the theological implications until later, the point of importance here is that, for Darwin, the metaphor of design was absolutely crucial. We return to the point made in the last chapter about selection and adaptation. Darwin did not think that eyes are simply thrown together. They function. They work. They exist in order that animals might see. They exhibit, and are explicable through, a final cause. Adaptation, "contrivance" as Darwin often called it, is at the heart of his theory. This is the problem that natural selection sets out to explain. You may think that Darwin overdoes adaptation. You may think that his metaphor is leading him astray. We shall encounter people who think precisely that. But that is beside the point at the moment. What is on the point is that Darwin's theory is impregnated with design thinking. It is, to use an eighteenth-century term, deeply respectful of "teleology," the idea that all phenomena in nature are determined by an overall design or purpose. The question we must therefore ask is about the status of this teleology. Is it just a help? Is it all a pragmatic matter, something that helps us to do our work, but that is essentially eliminable. We have a tree of life, but truly we could easily describe life's history without any help from botany or the Bible (where the tree first appears). We have the design metaphor but truly we could account for natural phenomena without the metaphor. Eyes are produced by natural selection, eyes see, end of discussion.

At least one major pre-Darwinian philosopher, Immanuel Kant, thought that teleology in biology cannot be eliminated. In the *Critique of Teleological Judgement* (1790), he stated flatly that the only way you can do biology is by assuming teleology or final cause. Without this assumption, you are stuck with questions that cannot be asked let alone answered.

> [Biologists] say that nothing in such forms of life is in vain, and they put the maxim on the same footing of validity as the fundamental principle of all natural science, that nothing happens by chance. They are, in fact, quite as unable to free themselves from this teleological principle as from that of general physical science. For just as the abandonment of the latter would leave them without any experience at all, so the abandonment of the former would leave them with no clue to assist their observation of a type of natural things that have once come to be thought under the conception of physical ends.
>
> (Kant 1928, 25)

Kant then went on to say that we simply must invoke something that he called a "regulative principle." He wrote: "Strictly speaking, we do not observe the *ends* in nature as designed. We only read this conception *into* the facts as a guide to judgement in its reflection upon the products of nature. Hence these ends are not given to us by the Object" (p. 53). And, spelling things out:

> All that is permissible for us men is the narrow formula: We cannot conceive or render intelligible to ourselves the finality that must be introduced as the basis even of our knowledge of the intrinsic possibility of many natural things, except by representing it, and, in general, the world, as the product of an intelligent cause – in short, of a God.
>
> (p. 53)

Now I do not want to get too bogged down in the history of philosophy or of ideas here. As it happens, I believe that Kant was an indirect influence on Darwin, through the great French comparative anatomist Georges Cuvier, although I believe that there were more immediate British influences. As it also happens, I believe that Kant was stuck in a conundrum from which he could not extricate himself. He thought that there could be no natural explanation of adaptation; he himself believed that God was responsible, but at the same time he thought it illegitimate to introduce God into science. However, these interesting issues are not my main point. This is that Kant is right to say that there is something peculiar, something distinctive, about biological science compared to the physical sciences. It has this central metaphor of design and it cannot be removed. Final cause is here to stay, and Charles Darwin reinforces this.

The point I am making is that, because of this essential metaphor of design, the very logic of evolutionary thinking is different from that of the physical sciences. Possibly you could get rid of the metaphor, but only by heroic surgery. You would not have the theory with which you started. But why feel the need of such surgery? The living world is different from the non-living world. Simply accept this fact and that it calls for different modes of explanation – not better, not worse, just different. Leave it at that.

Conclusion

Today we accept natural selection as the mechanism of evolution. The coming of Mendelian genetics made all the difference. Someone like

Huxley could have done a bit more with natural selection than he did. But he had other game to pursue. In this chapter, I have been more interested in realities than in hypotheticals, in what Darwin did rather than in what he or others might have done. Now let us move the clock forward and see whether Darwin did in fact pave the way for more to be done, for possibilities to be realized, for natural selection to come into its own. In the next three chapters we shall try to answer this question.

4

Neo-Darwinism

I turn now to look at developments in evolutionary thinking since Darwin. My coverage will be ahistorical, focusing not on the order in which discoveries were made and hypotheses proposed but on the order in which ideas and topics are presented in the *Origin of Species*. This means that I shall start with the empiricist *vera causa* principle and how selection measures up to this today.

Evidence for selection

The empiricist asks for evidence of natural selection at work – either direct evidence, or indirect evidence in the form of an analogy, namely artificial selection. At the time of writing the *Origin of Species*, Darwin had no direct evidence of selection. He strikes me as almost casual about the lack of such evidence. I suspect that he was so certain that he had technically satisfied the philosophers' criteria for *vera causa* status, he was not much bothered by failing to find selection visibly at work today. Perhaps he thought, also, that the workings of selection would be so slow it would be difficult to capture it in action. Not that Darwin was completely indifferent to such issues. The work of Henry Walter Bates on mimicry in Lepidoptera in the early 1860s showed that it was possible to find and test selection claims. Bates's work may have been no more substantial than a puff of smoke, to be blown away by the different aims of Huxley and his fellows, but Darwin appreciated it, and mentioned it in later editions of the *Origin* (albeit towards the end, and not where selection was first introduced).

There is another and curious class of cases in which close external resemblance does not depend on adaptation to similar habits of life, but has been gained for the sake of protection. I allude to the wonderful manner in which certain butterflies imitate, as first described by Mr. Bates, other and quite distinct species. This excellent observer has shown that in some districts of South America, where, for instance, an Ithomia abounds in gaudy swarms, another butterfly, namely, a Leptalis, is often found mingled in the same flock; and the latter so closely resembles the Ithomia in every shade and stripe of colour, and even in the shape of its wings, that Mr. Bates, with his eyes sharpened by collecting during eleven years, was, though always on his guard, continually deceived. When the mockers and the mocked are caught and compared, they are found to be very different in essential structure, and to belong not only to distinct genera, but often to distinct families.

> (Darwin 1959, 666–7; the first mention of Bates's work
> is in the fourth edition, 1866)

What is the cause of this phenomenon?

We are next led to enquire what reason can be assigned for certain butterflies and moths so often assuming the dress of another and quite distinct form; why, to the perplexity of naturalists, has nature condescended to the tricks of the stage? Mr. Bates has, no doubt, hit on the true explanation. The mocked forms, which always abound in numbers, must habitually escape destruction to a large extent, otherwise they could not exist in such swarms; and a large amount of evidence has now been collected, showing that they are distasteful to birds and other insect-devouring animals. The mocking forms, on the other hand, that inhabit the same district, are comparatively rare, and belong to rare groups; hence, they must suffer habitually from some danger, for otherwise, from the number of eggs laid by all butterflies, they would in three or four generations swarm over the whole country. Now if a member of one of these persecuted and rare groups were to assume a dress so like that of a well-protected species that it continually deceived the practised eyes of an entomologist, it would often deceive predaceous birds and insects, and thus often escape destruction.

> (pp. 667–8)

In other words, here we have an excellent illustration of natural selection.

There were other cases where naturalists would argue that selection may well have been operative, but even at the century's end they were not common. One of the most famous examples is that of Herman

C. Bumpus, who gathered up house sparrows exhausted, injured, or killed in a storm and measured them for evidence of natural selection at work. He argued that there was a significant difference between the survivors and those who died; that this difference was a matter of being closer to the norm (survivors) or further from it (losers); and that this was a case of selection in action. Too big, too small, or abnormal in some other way, and you are in trouble. (Today this is known as "stabilizing selection.")

> A possible instance of the operation of natural selection, through the process of the elimination of the unfit, was brought to our notice on February 1 of the present year (1898), when, after an uncommonly severe storm of snow, rain, and sleet, a number of English sparrows were brought to the Anatomical Laboratory of Brown University. Seventy-two of these birds revived; sixty-four perished; and it is the purpose of this lecture to show that the birds which perished, perished not through accident, but because they were physically disqualified, and that the birds which survived, survived because they possessed certain physical characteristics. These characters enabled them to withstand the intensity of this particular phase of selective elimination, and distinguish them from their more unfortunate companions.
>
> (Bumpus 1898, 209–10)

It was not until the 1930s and 1940s that selection studies really caught fire. If we fast forward to the present, we have instances too numerous to list (Endler 1986; Ruse 2006a). To start with human diseases and the effects of antibiotics and other medicines: selection is the reason why fighting things like venereal diseases needs ever higher doses of penicillin. Then go on to the whole area of agriculture, where human interventions cause changes that come about through selection. It is well known that a new pesticide is initially very effective against pests, but that the pests recoup and come back even more strongly than before (Committee on Strategies for the Management of Pesticide Resistant Pest Populations 1986). Some few members of the group have variations that enable them to withstand the attack of the poison on their systems, they are selected, and before long the whole population is protected. Consider topics like mimicry, started by Bates and now a subject of massive study by naturalists. And (as exhaustion sets in) move on to effects of climate and the like, and then to ecology and predation, and a host of similar areas.

The classic case of observed natural selection in action is that of the moths of England (Majerus 1998). Some of these have changed from rather mottled ("peppered") forms to having dark-colored wings, affording protection against predating birds, as their background trees have darkened through pollution. This fact had been noted even by the end of the nineteenth century.

> In our woods in the south the trunks are pale and the moth has a fair chance of escape, but put the peppered moth with its white ground colour on a black tree trunk and what would happen? It would . . . be very conspicuous and would fall prey to the first bird that spied it out. But some of these peppered moths have more black about them than others, and you can easily understand that the blacker they are the nearer they will be to the colour of the tree trunk, and the greater will become the difficulty of detecting them. So it really is; the paler ones the birds eat, the darker ones escape. But then if the parents are the darkest of their race, the children will tend to be like them, but inasmuch as the search by birds becomes keener, only the very blackest will be likely to escape. Year after year this has gone on, and selection has been carried to such an extent by nature that no real black and white peppered moths are found in these districts but only the black kind. This blackening we call melanism.
>
> (Tutt 1891)

In the 1950s, H. B. D. Kettlewell did massive studies on moths (published in toto in 1973), showing how they are differentially predated by birds, those that are peppered doing well on unpolluted trees, and those that are melanic doing well on polluted trees. Recently this work has been controversial because the American biblical literalists (more on these folk in a later chapter), knowing what a key piece of evidence it is for evolutionists, have seized on Kettlewell's work as fraudulent. This is simply not true. The work has been analyzed and reanalyzed, using statistical techniques far more powerful than he had in his grasp. The conclusion is unequivocal.

> Kettlewell's famous mark-release-capture experiments in Birmingham and Dorset provide strong evidence for the differences in fitness of the forms in polluted and unpolluted environments. Furthermore, his observation that live moths, which had been released on to tree trunks, were differentially preyed upon by birds provides evidence that the intensity of bird predation on the morphs varies according to habitat.
>
> (Majerus 1998, 126)

In addition, there have been many repeat experiments confirming Tutt's original hypotheses. Moreover, similar effects have been found in other organisms. And, to cap it all, one of Tutt's key predictions – that if and when pollution is tackled and the trees regain their old skins, the moths will reverse the trend and the norm will once again be speckled – turns out to be true.

For obvious reasons, many of the best-known cases of natural selection in action have involved responses to human intervention with nature – pollution, pesticides, and so forth – where an action often taken for medical or economic purposes seems to work well for a while and then becomes less and less effective as nature responds with its defenses. But not all such cases result from our actions. As an example of selection not caused by human activity one may cite a famous study by a husband-and-wife team, Peter and Rosemary Grant, of the finches on the Galapagos. They ringed all of the birds on one small islet (on average an islet will contain about 1,200 specimens), and were able to record changes in beak size tightly correlated to food supplies, in turn a function of changes in the climate. When there is drought, the only foodstuffs available tend to be hard and big – nuts and cactus and the like – and the big-billed finches are favored. When it rains, there is an abundance of all kinds of foodstuff, and smaller-billed finches are at least as efficient as their species mates, if not more so. Selection occurs and selection works (Grant 1986; Grant and Grant 1989).

What about artificial selection in action? Again today there are huge numbers of studies on which one can draw. Justly famous is a very long-term study (it was started in 1896) at the University of Illinois on oil content in corn (or maize). At the start of the study, the corn had an oil content of around 4–6 percent. By the end of the study, after selection, the corn was up to an oil content of about 16 percent, a threefold increase (Dudley 1977). Other more recent studies have tried to replicate conditions that might have led to changes. One intriguing example of selection is that of the little fruitfly, which is very sensitive to alcohol. Normally, even low concentrations prove fatal. However, around wineries the fruitflies not only are alcohol-tolerant but positively thrive on the stuff. Is this magic, something put in place by the Gods to promote conviviality among the insects? More likely there was fairly severe selection in the fruitfly populations around wineries: minute variations in alcohol tolerance were picked up and selected, until today's boozing fruitflies were produced.

Wineries have proved to be popular study sites for Drosophila researchers in many parts of the world. The interior of a winery cellar and the neighborhood of wine fermentation vats are good places to find high ethanol environments and to test hypotheses developed from laboratory data and from studies on natural populations in orchards. The distribution of Drosophila species inside and outside wineries is in uniform accord with expectations; D. melanogaster [a species with high alcohol tolerance] is found at much higher frequencies inside wineries than is the closely related but relatively alcohol-sensitive species D. simulans.

(Chambers 1988, 69)

Geneticist Francisco Ayala and his students put the selection hypothesis to the test, upping the alcohol concentration for caged groups of flies and breeding from the survivors. Before long, they had produced from alcohol-sensitive flies precisely those alcohol-tolerant flies found in the wild around sources of alcohol (McDonald, Chambers, David, and Ayala 1977).

Now, what about Thomas Henry Huxley's question, what about speciation? Can artificial selection separate interbreeding groups? We can certainly produce sterility in plants fairly easily. Sometimes interbreeding causes sterile offspring that are unable to breed with members of either parent group. For instance, you can cross a cabbage with a radish and get an offspring plant that can reproduce with neither parent group; unlike a mule, which is sterile all around, the offspring can be mated with its own group and proves fertile (though unfortunately, in this case, the result is not much good, for it has the root of a cabbage and the leaves of a radish). Can you get a reproductive breakdown in the case of members of one species selected apart? Again, the answer is yes. One experiment on corn (Paterniani 1969) started with yellow and white varieties. In each generation, (artificial) selection favored those yellow and white members of the group that did not breed with other-colored members. In other words, selection was for white–white breeders and yellow–yellow breeders. In only five years, barriers were being set up. White–yellow breeders dropped from 35.8 percent to 4.9 percent, yellow–white breeders from 46.7 percent to 3.4 percent. There was no particular magic about any of this. Selection worked on the white variety to start it flowering at an earlier time, and on the yellow to flower at a later time. Thus, by the fifth generation, there was less chance of the populations producing hybrids. Obviously, making the analogy to the natural world, it is easy to think of natural conditions where climate or other factors might separate groups in this way.

We can leave the discussion here. Has someone actually turned a cow into a horse? No, obviously not. Apart from anything else, you would not expect to turn a cow into a horse, because the one was not the ancestor of the other. Even if it were possible, there is not the time that would be needed. But, with various degrees of human involvement, evolutionists have found countless examples of selection working as one expects; and in the absence of argument to the contrary, there is no reason to think that it should not work long term and cause biological differences of the greatest and most profound kinds. There is lots of evidence to support an empiricist *vera causa* approach to natural selection.

Mendelian genetics

We come now to the modern-day equivalent of those parts of the *Origin* that introduce natural selection and try to show how it works – what you might properly call the heart of the whole theory. In order to look at how it is conceived and presented today, we need first to turn to the other great figure in the history of evolutionary thought, Gregor Mendel, a German-speaking Augustinian monk, living in the Austro-Hungarian Empire, in an area that today is part of the Czech Republic. Fascinated by questions of biology, in the 1860s Mendel performed what we now appreciate as ground-breaking studies on the nature of variation. At the time no one recognized them as such, and for the last period of his life he gave up science and, becoming abbot of his monastery, channeled his labors into administration. It was not until the beginning of the twentieth century that others working independently hit on the same ideas as Mendel and then, searching back in the literature, found that they had been trumped by a virtually unknown cleric.

Much energy has gone into speculations about what might have happened had Mendel been appreciated by his contemporaries for the genius that he was. It is hard to say, but one should not assume automatically that people would have seen the full relevance of his work. Mendel himself seems not to have done so. He was most interested in the problems of evolution, reading and annotating a German translation of the *Origin*. But there is no sign that he thought of his own work as filling the gaps. (Darwin did not read or know of Mendel, although he did own a book that referred to Mendel's work.) In fact, even after Mendel's work was rediscovered, it was a long and hard grind before Darwin's and Mendel's work was seen as complementary, giving different parts of the overall picture, rather than as contradictory, giving rival theories of evolution. By the beginning of the twentieth century, even (in fact, especially) those few people who thought that there might really be something to selection, were not that receptive to the new findings about heredity (Provine 1971; Ruse 1996). Personally and intellectually the partisans of Darwin and Mendel were at odds, differing mainly over the nature of variation in populations. The Darwinians – the English Darwinians, that is – known because of their use of mathematics as the "biometricians," thought that variation is minute and that selection is important. The Mendelians, mainly because they worked on easily spotted variations, thought that variation is large and selection is unimportant: they were saltationists.

There was a divide, and some never saw the connections. Slowly, however, the full picture started to emerge. Let us begin with the new theory of heredity, then move to the synthesis with natural selection. In order to start the discussion, we need to recognize that the new theory of heredity owed as much to cytology, studies of the cell, as it did to breeding experiments, the part that feeds directly into genetics. Focusing on the developed theory – the so-called "classical theory of the gene" – which owed much or most to the American biologist Thomas Hunt Morgan, who worked with a team of students at Columbia University in New York in the second decade of the twentieth century, we therefore start with the cell, that self-enclosed unit from which living beings are built. In the center of the cell is the nucleus, and this contains a number of string-like entities, the chromosomes. In sexual organisms, these chromosomes are paired, and it is on the chromosomes that the units of heredity, the genes, are to be found. In the classical theory, the genes are basically unanalyzed entities, although now – thanks to the work in 1953 of the molecular biologists Francis Crick and James Watson – we know that the genes are long molecules of deoxyribonucleic acid, DNA. (Sometimes the genes are another nucleic acid, ribonucleic acid, RNA. In this account, I am ignoring exceptions, unless absolutely necessary for exposition.)

The set of genes that an organism has is known as the "genotype," and this is contrasted with the physical body of the organism, the "phenotype." With some special exceptions, the collective set of genes of an organism is repeated from cell to cell throughout the body. The development of the organism, therefore, is a matter of activation of different genes to make different parts at different times. It is not a matter of different sets of genes. The key exceptions to the paired chromosome content of cells (called "diploid cells") are the sex cells (gametes). These carry only a half set of chromosomes (that is, they are haploid), one half in each of the pair of cells. In fertilization, one haploid set from each parent (sperm, ovum) is joined to make a cell with a new diploid set (zygote); this then multiplies to produce the new individual. The production of diploid cells from diploid cells is called "mitosis." In the production of sex cells (from diploid cells), the chromosome sets are halved in a process known as "meiosis."

We are now ready to introduce Mendel's first law (remembering that we can do it in the light of our knowledge of the cell, something he did not have). The double chromosomes in diploid cells are physically identical – the same length and so forth. (The exceptions are the sex chromosomes: these differ and one is generally shorter than the other.) Genes occupy particular places on the chromosomes, known as "loci." A particular kind of gene will always occupy the same place (locus) on a chromosome, and the set of gene variants that can occupy a particular locus on the chromosomes are known as "alleles." In diploid cells, that is with two half sets of chromosomes, there will consequently be matched allele-pairs at any particular locus. If the matched alleles are identical, the organism is called "homozygous" (with respect to that locus) and if different, "heterozygous" (with respect to that locus). Mendel's first law tells us that in the zygote there is one chromosome set from each parent. That is, they are paired. Hence, at any particular locus on the chromosomes, one of the allele-pairs is from one parent and the other from the other parent. And – this is the crucial bit – it is purely chance as to which allele one gets from one parent and which from the other. It is not that the transmission is uncaused, but that the law specifies that it is equiprobable which gene from a pair of parental alleles will be transmitted.

Thus if you think of one locus, with four possible alleles, A, B, C, and D, and you crossed a heterozygous AB with a heterozygous CD

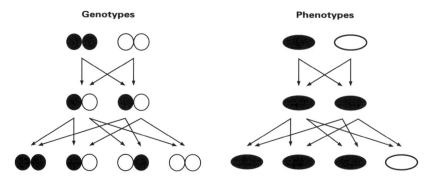

Figure 4.1 Mendel's first law: Black (allele A) is dominant over White (allele a). The genotypes are AA, two Aa, aa. The phenotypes are three Black, one White

(AB × CD), you would expect (on average) that one quarter of the offspring to be AC, one quarter AD, one quarter BC, and one quarter BD. What can make this law so difficult to discern, when you are just working (as was Mendel) with the results of breeding, is that sometimes one allele will conceal the effects of another. If you have two alleles, A and a, and the homozygotes (AA and aa) are different, but the heterozygote (Aa) looks like one of the homozygotes (say AA), then the one allele (A) is said to be dominant over the other (a) which is said to be recessive. In this case, if you bred two heterozygotes (Aa × Aa), you would expect to get one quarter AA, one half Aa, and one quarter aa (since Aa is the same as aA). But since the Aa looks like (has the same phenotype as) AA, the phenotypes would appear to be three-quarters AA and one quarter aa (Figure 4.1).

Mendel's second law tells us that what happens at one locus does not have any bearing on what happens at another locus. In other words, the transmission of alleles at one locus tells us nothing about the transmission of alleles at another locus. Early it was realized that this law has a major exception, namely at loci on the same chromosome. You might think that genes on some particular chromosome must always go (or not go) together. This is not quite true. During meiosis the paired chromosomes line up, break along their lengths (in the same spots), and exchange pieces with their mates. This is known as "crossing over" and ensures that even the genes on the same chromosome are subjected to some shuffling during reproduction. Obviously, the closer together two loci are on a chromosome, the less chance there will be that a break occurs between them – that is, the more chance that they will remain linked together. This insight led Morgan's group to the triumphant mapping of the positions of genes along chromosomes. Interestingly Mendel's work (which was on pea plants) reported on genes only on different chromosomes – where his second law held without exception. This has led some to suspect that Mendel was selectively choosing which features to report and which not. There is certainly some strong evidence that his reported distributions after breeding were far too close to expectation to have happened in the normal course of events. Someone was cleaning up or massaging the data.

One final point and then we can turn to evolutionary questions. We have so far seen that the gene is the unit of function: it is this that brings about (or carries the information to make) the physical body. We have also seen that the gene is the unit of heredity: it is the gene that carries the information from one generation to the next. It is these two aspects of the gene that we are talking about if, for instance, we are trying to explain why sometimes brown-eyed parents will have blue-eyed children but that blue-eyed parents will never have brown-eyed children. If A is dominant over a, and AA produces brown-eyed people and aa produces blue-eyed people, then two heterozygotes (Aa) who themselves are brown-eyed should have blue-eyed children one quarter of the time, but two blue-eyed people (aa) can only produce similar homozygotes (aa), who will likewise be blue-eyed.

There is a third aspect to the gene in classical gene theory. It is also the unit of change. Sometimes genes move from one form to another. In other words, one allele changes to another, different, allele. This is known as mutation. Mutation is not uncaused – one of Morgan's students, H. J. Muller, won a Nobel Prize for his work showing that radiation is one of the major causes of such change – but within the system it is simply taken as a given. It is random in at least two senses. First, it does not occur to need: if an organism would benefit by some change that engendered camouflage, mutation is not directed to provide it. Second, mutation is quantifiable only over groups: you cannot, for instance, say that this organism with allele A will mutate to provide a gamete with allele a; at best, all you can say is that X percent of A alleles in a population will mutate to a in any one generation.

Population genetics

Natural selection is a mechanism that can work only in a group situation. To select is to pick from a range of options. If selection and genetics were to be blended, it was vital that genetics be stretched out and applied to populations. In principle this is fairly easy to do, although before it could be done it was necessary to shift the focus of evolutionary theory somewhat, from the individual organism to the gene – in the language introduced above, from the phenotype to the genotype. Darwin's theory in the *Origin* talks about finches and ants and wolves and turnips and pigeons and on and on. In modern theory, these do not cease to exist, but now one reconceptualizes things and recognizes that with the coming of genetics what matters in heritable transmission is not so much the individual but the units of heredity, the genes. If you like, the bookkeeping is now done at the level of the genes rather than the level of the individual. This means that natural selection is a matter of the differential copying (or reproduction) of the genes rather than of individuals. Genes that produce more copies for the next generation (through the physical bodies that they produce) are fitter than genes that do not. There is good reason for this shift of focus from individual organisms to the genes. Other than because of spontaneous changes (mutation), the genes – or rather, copies of the genes – stay the same from generation to generation, whereas, because of the randomizing effect of Mendel's laws, individuals (clones apart) are unique in each generation. Conceptually it is much easier for a scientist to follow genes than organisms.

How is the crucial Mendelian law of transmission (the first law) to be generalized? You must think of a population (conceived now as a population of genes not organisms) and, to get things going (meaning, to build a model starting at the simplest level), you must focus on the alleles at one locus only. Suppose there are just two kinds of gene at a locus (in other words, in the population there are two alleles): A and a. Suppose also that the ratio of A to a is $p : q$ (meaning that $p + q = 1$). Then simple mathematics shows that (outside forces being absent) the original ratios of genes in a population will remain stable (in other words, the ratio will remain at $p : q$). Moreover, after the first generation, no matter what the original distributions of the genes in a population, they will always appear in individuals in determinable fixed ratios. We have the possibilities of three genotypes: homozygote AA, heterozygote Aa, homozygote aa. The rule determining the distribution of genotypes is called after its discoverers: the Hardy–Weinberg Law. It states that the equilibrium distribution is:

$$p^2 \text{ AA} : 2pq \text{ Aa} : q^2 \text{ aa}$$

The theory of heredity generalized like this is known as "population genetics." It is at the heart of modern evolutionary thinking, and at the heart of population genetics – the fundamental premise or axiom – is the Hardy–Weinberg law. Notice that it is an equilibrium law and functions much like Newton's first law of motion – if nothing happens, then nothing happens. It is only if external forces impinge that change happens. In neither the Newtonian case nor the biological case are we dealing with a tautology. Empirically, the ratios could always go to the larger proportion (that is, where $p = 1$ and $q = 0$ we end up with nothing but A). Early geneticists thought just this, and it took the Cambridge mathematician G. H. Hardy – no biologist he – to put them right. But in their respective systems, the equilibrium laws are absolutely crucial because they provide a background of stability. You can now start to introduce forces for change.

In the biological case, you can talk about the migration of individuals into and out of populations – or rather the migration of genes into and out of populations – thus changing gene ratios. You can talk about mutation, which likewise changes ratios. (For instance, in any generation a certain number of A alleles might turn to a, and conversely a different number of a to A.) And then there is selection. Notice that (unlike Darwin's theory) selection is now introduced after heredity comes in, and indeed is only considered effective if it can disrupt the background of genetic stability. This does not mean that selection is any less important, but it has a slightly different logical position in the neo-Darwinian theory. For instance, you might suppose that in any population a certain percentage of organisms, or rather of genes, are going to be wiped out by selection (the selective force or selection pressure), and then, by factoring this into the Hardy–Weinberg law, you can show what proportions or percentages of genes will be present in the next generation.

Conceptually, you would not think of selection as happening against a certain gene, but rather as against the individual. Translated at the genetic level this means the genotype, and if you are considering just one locus it would be a matter of selection against one or other of the two homozygotes and the heterozygote. So let us suppose that one allele is dominant over another allele. And let us suppose that the homozygote of the recessive gene is less fit than the other two genotypes. The measure of fitness is written as ω. Generally, you speak of a perfectly fit organism as having a measure of unity and of a totally unfit organism as having a zero measure (no offspring at all). So we can speak of a coefficient of selection (symbolized as s) as something that detracts from the recessive-gene homozygote. We can write:

Genotype	AA	Aa	a
Fitness (ω)	1	1	$1 - s$

Assuming the initial distribution of A to a is $p : q$ $(p + q = 1)$, then we can draw up the matrix shown in Table 4.1, given that the original distribution was in Hardy–Weinberg equilibrium and would stay that way were selection not interfering.

Table 4.1 Selection changes Hardy–Weinberg ratios

	Genotypes			Total population
	AA	Aa	aa	
Initial zygote frequency	p^2	$2pq$	q^2	1
Fitness (ω)	1	1	$1 - s$	
Zygote proportions after selection	p^2	$2pq$	$q^2(1 - s)$	$1 - sq^2$
Zygote frequencies after selection	$\dfrac{p^2}{1 - sq^2}$	$\dfrac{2pq}{1 - sq^2}$	$\dfrac{q^2(1 - s)}{1 - sq^2}$	1

Now let's see what happens to A. How does it increase in the next population with respect to a? In other words, what is the value of Δp, the increment in the value of p, the original frequency of A? Set X as the new frequency and, remembering that there are two As in each AA, and one A in each Aa, we get:

$$X = \frac{2p^2 + 2pq}{2(1 - sq^2)} = \frac{p(p + q)}{1 - sq^2} = \frac{p}{1 - sq^2}$$

Hence:

$$\Delta p = X - p = \frac{p - p(1 - sq^2)}{1 - sq^2} = \frac{spq^2}{1 - sq^2}$$

What does this all mean exactly? Well, suppose that you have a population where the two alleles A and a start equal ($p_0 = q_0 = 0.5$). Think of a range of possibilities, where the selection coefficient ranges from s = 0.01 (the homozygote has a 1 percent disadvantage) to s = 1 (the allele a is lethal, for the homozygote always fails to reproduce). Then:

Selection coefficient, s	0.01	0.02	0.1	0.5	1.0
Frequency of A, p_1	0.50125	0.5025	0.5128	0.547	0.67

And what does this mean? Two things, really. First, even with the weakest selection, a is going to decline. No surprise here. But second, and a real surprise here, even with the strongest selection – the homozygote aa always dies without reproducing – the a allele will linger on and on in the population. If we start with a frequency of 50 percent ($q = 0.5$), it takes only eight generations to get it down to 10 percent. It takes a hundred generations to get it down to 1 percent, and a thousand generations to get it down to 0.1 percent. This means that in a large, human-size population, especially one that is relatively slow breeding, you are likely to have aa individuals for virtually indefinite years to come. Utopian plans of eugenics, hoping to eliminate all genetic disease by preventing those affected from breeding, are almost doomed to failure.

The point is made. Theorizing like this has replaced the original deductions that Darwin gives in the third and fourth chapters of the *Origin*. Note that our calculations are more formal than Darwin's, which is what you would expect as a science matures, and (although I am somewhat dubious about the significance of this), for those who care about such things, we can now say that we are dealing with entities (genes) that are not observed directly. We are nevertheless still absolutely and completely dealing with a Darwinian theory. Selection is the big force that changes things.

Sickle-cell anemia

Let us now see how evolutionists use and apply their core theory of population genetics, the synthesis of Darwinian selection and Mendelian genetics. Few today are ignorant of the genetic disease sickle-cell anemia, which afflicts African Americans. In that population, one often encounters children who suffer and (at least, without heroic medical intervention) die young from a heritable form of anemia. They simply do not have a high enough red-blood-cell count, or rather those red blood cells that they have are deformed and low functioning, taking on an outline that is more sickle shaped (hence the name) than round (the usual shape). The anemia has been studied in great detail and its causes are known fully. It is caused by a recessive gene that brings on the collapse of the cells in homozygotes. Heterozygotes do not show the same symptoms, although there have long been debates about whether they are in any sense anemic or susceptible to anemia.

Why should this anemia persist since it is so dreadful? Why have the genes not been selected out? If we go back to Africa, from where the genes originated, the answer soon becomes apparent (Allison 1954a, 1954b). The gene flourishes, if that is not too awful a term, in those areas of the continent where malaria is endemic. Look first at the maps of Uganda in 1949, comparing the distributions of the sickle-cell gene with the distributions of the types of malaria, *Plasmodium falciparum* (Figure 4.2). The isomorphism is way too close to be mere

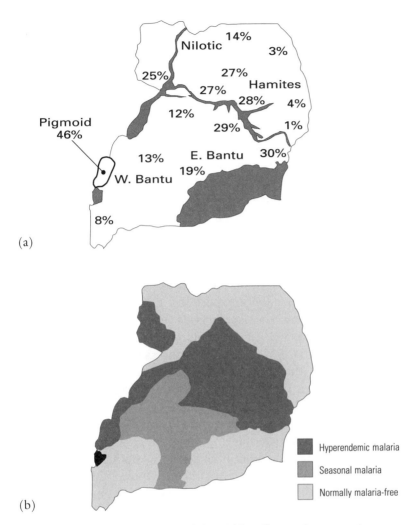

Figure 4.2 Distributions (a) of the sickle-cell gene in proportion to normal genes among the indigenous peoples of Uganda, and (b) of types of malaria in Uganda (*c*.1949)

coincidence, and connection is even closer when you look at the world-wide native distribution of the sickle-cell gene compared to the incidence of malaria. There surely has to be some causal relationship, and evolutionary biologists – Darwinian biologists – have been able to show what it is. Heterozygotes for the sickle-cell gene – that is, people with one normal gene and one sickle-cell gene – have a natural immunity to malaria not possessed by those without any sickle-cell genes. It is all quite simple, in fact. *Plasmodium falciparum* infects the red blood cells. The cells of those who lack sickle-cell genes remain more or less functional, and hence the malaria spreads through the body. The cells of heterozygotes, however, tend to collapse when infected, and at once the white blood cells remove them; so the infection is removed and cannot spread.

In other words, the patterns of sickle-cell genes can be shown to be a function of the protection that they give against malaria, and the distribution is explained in a good Darwinian fashion. We can go a bit further with our evolutionary explanation, however. How can it be that the benefits of having one sickle-cell gene are enough to balance out the drawbacks of having two sickle-cell genes? Remember, by Mendel's law, if two carriers of the sickle-cell gene have four children, then on average you expect that one of the children will have full-blown anemia because it is a sickle-cell homozygote – and, for that matter, one of the children will have no protection at all against malaria, because it is a non-sickle-cell homozygote. Can we show that the good selective effects of two heterozygotes (expected on average) will balance the bad selective effects of the two homozygotes? Well, yes, as a matter of fact we can. (Notice, incidentally, that here we are going to have a paradigmatic example of building a model and then applying it to nature.) Suppose that the heterozygote is fitter than either homozygote. This means we put the heterozygote at unity. Suppose, also, selective forces (call them s and t) against the two homozygotes. Then we can draw up the following matrix:

Genotype	AA	Aa	a
Fitness (ω)	$1 - s$	1	$1 - t$

Then, modifying our original table, we get the results shown in Table 4.2, and can calculate the rate of change as follows:

$$\Delta p = \frac{pq(tq - sp)}{1 - sq^2 - tq^2}$$

Table 4.2 Hardy–Weinberg ratios with selection against homozygotes

	Genotypes			Total population
	AA	Aa	aa	
Initial zygote frequency	p^2	$2pq$	q^2	1
Fitness (ω)	$1-s$	1	$1-t$	
Zygote proportions after selection	$p^2(1-s)$	$2pq$	$q^2(1-t)$	$1-sp^2-tq^2$
Zygote frequencies after selection	$\dfrac{p^2}{1-sp^2-tq^2}$	$\dfrac{2pq}{1-sp^2-tq^2}$	$\dfrac{q^2(1-s)}{1-sp^2-tq^2}$	1

Is there ever a situation where the various selective pressures "balance" each other out? In other words, do we ever get a situation where the change of gene ratio is zero? Putting $\Delta p = 0$, we get:

$$p = \frac{t}{s+t}$$

In this sort of situation, we have both genes persisting in each generation because the good effects of the heterozygotes balance the bad effects of the homozygotes. Finally, we can plug in some figures to show that the ratios of genes fit almost exactly what the models predict. In the sickle-cell example, assume that one in five die from malaria because they carry no sickle-cell allele (and all of the sickle-cell cases die of anemia). This gives you about three children in a hundred who have sickle-cell anemia, which is a correct prediction. All in all, we have a triumph of Darwinian reasoning.

Molecular biology

As I prepare to bring this chapter to an end, and get ready to open the next on the wide range of topics that Darwin subsumed within evolution through natural selection, the reader might have one justifiable (somewhat negative) comment. Thus far, the discussion has been very classical. Mention has been made of the great molecular revolution in biology, started by the Watson and Crick discovery of the structure of the DNA molecule. But nothing has been made of this or of the plethora of subsequent studies, culminating most recently in the mapping of the human genome. Is the suggestion that today's neo-Darwinian evolutionary theory is indifferent to everything that has happened at the molecular level in the past fifty years?

Simply, the answer is: Not at all! It is true that, in the early years, you could have cut the tension between the molecular biologists and the evolutionary biologists with a knife, or more likely you would have needed a chain saw. The molecular biologists were all-conquering – grants, students, prizes, adulation – and had no feel at all for the organism. They referred contemptuously to regular biologists as "stamp collectors." In return evolutionary biologists hated the young upstarts who were getting all of the goodies, and reacted with scorn to people who had to blast everything to smithereens before study could begin. More than one eminent evolutionary biologist turned to philosophy to prove that in the study of organisms the molecules are irrelevant and misleading. The big bugbear was "reduction," whatever that meant, and here was a stake to drive through the hearts of the molecule men – assuming, that is, that they had hearts in the first place. Having spent several pages running down molecular biology, eminent evolutionist Ernst Mayr wrote:

> It has never been demonstrated that reductionism works, so to speak, upward. To be sure, most of the phenomena of functional biology can be dissected into physical-chemical components, but I am not aware of a single biological discovery that was due to the procedure of putting components at the lower level of integration together to achieve novel insight at a higher level of integration. No molecular biologist has ever found it particularly helpful to work with elementary particles.
>
> In other words, it is futile to argue whether reductionism is wrong or right. But this one can say, that it is heuristically a very poor approach. Contrary to the claims of its devotees, it rarely leads to new insights at higher levels of integration and is just about the worst conceivable approach to an understanding of complex systems. It is a vacuous method of explanation.
>
> (Mayr 1969a, 128)

Wonderful stuff, but beneath the rhetoric is there any substance? I will answer three questions. First: What about the relationship

Neo-Darwinism

between Mendelian genetics and molecular genetics? Much philosophical ink has been spilt on this issue, and I am one of the offenders. Casting questions in traditional philosophical terms, the big issue is one of continuity. Everyone agrees that molecular genetics has replaced Mendelian genetics and that it has grown into something much more powerful. Was it a case of simply shoving Mendelian genetics to one side – nice but wrong, just like phlogiston theory, which was elbowed out by Lavoisier's chemistry – or did molecular genetics in some way absorb Mendelian genetics into itself? One meaning of reduction ("theoretical reduction") refers to this second option, and is usually defined technically in terms of deduction: an older theory is reduced to a newer theory if the older theory can be shown to be a deductive consequence of the newer theory.

You can see that this kind of definition rather implies that the two theories are hypothetico-deductive systems, and this is perhaps a reason why many of the bright young people today rather sniff at the possibility of reduction (Kitcher 2003). For myself, I will continue to adopt the tolerant – some might say flabby – attitude that I have taken on other issues. If you are looking for a tight deduction between the two genetics, then you will be disappointed. For instance, the Mendelian gene is considered one indivisible unit. The DNA molecule can itself be divided during crossing over. So they contradict, and one cannot be the consequence of the other. (The full story is a bit more complicated than this, because Mendelians knew that there was sometimes something suspect about the indivisibility assumption. Something known as the "position effect" challenged it. But, for crude discussion, the partial picture will do here.) Having said this, it is just plain silly to suggest that the DNA molecule brought everything that went before crashing down in flames, like Berlin in 1945. By and large, the earlier genetics was absorbed pretty readily and easily into the new genetics, and life went on much as before. The classical gene is a chunk of DNA and that's all there is to it. For this reason, the discussion in this chapter that casts things in terms of Mendelian genes is far from outdated. It works equally well whether you think the gene is some special biological unit or substance or whether you think it is a macromolecule. The problems and explanations are the same.

Second: Did the coming of the molecules in any sense help evolutionary studies? The answer is that they most certainly did. Genetic fingerprinting, the technique that lets you compare the DNA molecules of different organisms, has been almost as powerful and fruitful as computers. Time and again evolutionists want to check paternity. This was simply impossible, or dreadfully crude, until one was able to get unambiguous results by looking right down at the molecules. English ornithologist Nicholas Davies (1990) has done wonderful work on the little, drab bird known as the dunnock (the hedge sparrow). For something so inconspicuous, it has a sex life of stupendous variety, right out of a girlie magazine – there is polygyny (one male, several females), monogamy (one male, one female), polyandry (several males, one female), and something primly referred to as polygyandry, better known as group sex (several males, several females). Males help to feed the young and Davies wanted to know if the males apportion their effort according to their chances of being the father. With genetic finger-printing, he was able to determine every father for every child and draw his conclusions. The answer was a Darwinian triumph. By measuring the time that the males actually spent on a nest, Davies determined by this means which bird was father of which chick 143 times out of 144. Everyone does what is in their genetic interests, and not one bit more. Obviously the birds do not know about their genes. They follow fixed rules of behavior that give Darwinian results.

At a more theoretical level, one of the big questions asked by evolutionists since Darwin is exactly how much variation occurs in any population. In the 1950s, there was a major debate between H. J. Muller and the Russian-born American geneticist Theodosius Dobzhansky on this very issue, Muller thinking that most organisms are similar and Dobzhansky thinking that variation is the norm. (In the language of genetics, Muller thought that most of the alleles at some locus are identical, and Dobzhansky thought that they are varied.) Population geneticist Richard Lewontin (1974) developed the technique of gel electrophoresis to test the variations in an organism, and to the delight of Dobzhansky (Lewontin was his student) was able to show clearly that massive variation is always present in natural populations. One could hear the muffled cheers from the stones beneath the nave of Westminster Abbey.

Third: Has the molecular revolution actually made any difference to the theory of modern evolutionary thinking? In some respects it has. The two prominent founding theoreticians of population genetics were Ronald A. Fisher (1930) in England and Sewall Wright (1931, 1932) in America. They agreed on the sums, but differed very much on interpretation. Fisher, a fanatical Darwinian, saw natural selection as the force behind everything. Large populations get new variations thanks to mutation. If the new forms are better than the old, then they will spread because of selection. If not, then they will not. Wright had a more complex position. He believed that large groups get fragmented. Then he showed that in small groups, even if selection is operating, the effects of chance can be the determining factor. Because of the vagaries of breeding, it could be that less fit organisms get established in a group rather than the fitter. (This he called "genetic drift.") New features can be formed this way, and then, if the small groups connect up, in the larger group the hitherto inadaptive features might become adaptive and spread. There is thus a significant random factor built into what Wright called his "shifting balance theory of evolution."

Very successful was a powerful metaphor that Wright introduced to explain his theory. He supposed that the genes lie on an "adaptive landscape," and generally they move up and down hills according to their reproductive success. Translating Wright's shifting balance theory into this kind of language, the key notion is that sometimes genetic drift intervenes and a gene stuck on one hill can move down into the valley and up the slopes of a higher hill thanks to pure chance (Figure 4.3). Today, the metaphor of a landscape seems set to endure, but (although it does have its supporters) by and large genetic drift has not fared all that well. Time and again, phenomena that were thought to be a function of drift have been shown to be tightly controlled by natural selection. A paradigm instance occurred in the 1950s, when some young English researchers looked at patterns on the shells of snails (Cain 1954; Sheppard 1958). What had been assumed to be without function, just a result of drift, was shown to be governed by selection. It turns out that the shell patterns offer significant camouflage advantages, and that the snail's major predators, thrushes, trigger strong selective forces. Snails without the right patterns get seen and eaten. More than this: there is no one ideal pattern; the pattern that will best protect the snail depends on the background and habitat – woodland, meadow, ditch, and so forth. According to the location, selection promotes the best shell-color defense.

However, one place where drift does seem to be well established is down at the molecular level, where the forces of selection cannot be felt. Molecular biology has shown that there is a lot of redundancy and non-working (junk) DNA, and the Japanese theoretical biologist

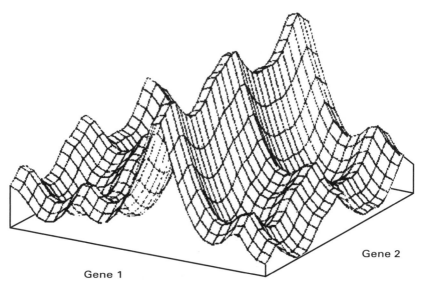

Gene 1

Gene 2

Figure 4.3 Adaptive landscape (after Wright 1931, 1932)

Motoo Kimura (1983) was able to show that, at this level, molecular variants drift from one form to another. This finding has been exploited by evolutionists, who argue that the rates of drift give one a way of making estimates of the time that has gone by since major events occurred. If a species splits into two, then the differences between them today, correlated with the rates of change, yield a powerful "molecular clock." I shall say more about this in the next chapter. The point here is that molecular biology has opened a whole new window into the world of organisms: non-Darwinian evolution (that is, non-selective evolution) down at the level of the DNA. It hardly needs saying that this is an extension of Darwin's thinking and not a refutation. Kimura was an ardent Darwinian. The *Origin* is not about this kind of phenomenon at all. And no more could it be. Until the discovery of the double helix, such a theory was not possible. But now it is.

Enough has been said. The molecular revolution is now welcomed by Darwinian evolutionists. As we shall see in the next chapters, the easy problems having been solved, the molecular biologists are themselves turning to the issues of evolution. They now realize that there is a lot more to the study of whole organisms than stamp collecting.

Neo-Darwinism

5

The Consilience: One

We turn now to the other side (the rationalist side) of the *vera causa* question, and see how the consilience of inductions fares in modern evolutionary theory. I shall follow Darwin's order of presentation, but shall begin with something entirely missing from the *Origin*.

The origin of life

In one of my favorite Sherlock Holmes stories, *Silver Blaze*, the great detective is asked about clues that might have been missed.

> "Is there any point to which you would wish to draw my attention?"
> "To the curious incident of the dog in the night-time."
> "The dog did nothing in the night-time."
> "That was the curious incident," remarked Sherlock Holmes.

I always think of this when I read the *Origin of Species*. In Conan Doyle's story, Holmes realized that the silent dog was significant because it should have been barking when the intruder was at work. Holmes reasoned that whoever it was who had done the evil deed (of stealing the favorite for the Wessex Cup and killing the horse's trainer) must have been known to the dog. The crime was an inside job. (For those who are so culturally deprived as not to know the story, it turns out that the trainer himself was trying to nobble the horse and was killed when the animal lashed out.)

What is the dog in the *Origin of Species* and why does it not bark? The dog is the discussion of the ultimate origins of life, from non-life. Before the *Origin*, everyone assumed automatically that if you are into the evolution business, then you must discuss where life came from in the first place (Farley 1977; Fry 2000). Lamarck came up with all kinds of speculations about worms being "spontaneously generated" from mud by the action of heat and electricity and so forth. This is an idea that goes back to Aristotle and can be found in just about every other evolutionist before Charles Darwin, including Grandfather Erasmus.

> Then, whilst the sea at their coeval birth,
> Surge over surge, involved the shoreless earth;
> Nursed by warm sun-beams in primeval caves
> Organic life began beneath the waves . . .
> Hence without parent by spontaneous birth,
> Rise the first specks of animated earth.
>
> (Darwin 1803)

Robert Chambers in *Vestiges* (1844) gave the reader all sorts of information about insects appearing in the course of experiments on electricity and about how the frost ferns found on windows in the middle of winter look so much like real ferns. The *Naturphilosophen* simply assumed that everything was part of some great overall plan or picture, and so they had no problem with seeing life emerging from non-life. After the *Origin* there was continued discussion of the topic. Huxley really thought that the most primitive life forms had been found, with evidence of their non-living origins. So why did the dog not bark? Why was Darwin silent? Why was his only comment that life started from some one or a few forms?

It was not because Darwin was not interested in the topic. In fact, in one of his letters written some time after the *Origin*, he speculated on life forming in some nice, warm little pond. Rather it was because he knew that ideas about spontaneous generation were coming crashing down – Louis Pasteur in France was doing his celebrated experiments showing that life did not come from non-life – and Darwin realized that if he opened up that particular can of worms, it would never again be closed. Discussion about evolution and natural selection would be lost in the overall debate. So, on the principle that

The Consilience: One

saying nothing is sometimes the best strategy, that is precisely what Darwin did. And very successful it proved, because people did grab onto evolution, even if they were less than enthused about selection. But of course the origin of life was a topic that would not go away, and today no discussion of evolution would be complete without it. It has also become a focus of philosophical debate, with those who are not very keen on science and its claims about origins holding up this particular issue as an impassable barrier to the complete success of any naturalistic (that is lawbound) explanation. Notre Dame philosopher Alvin Plantinga writes of any claim that life emerged naturally from non-life: "This seems to me for the most part mere arrogant bluster; given our present state of knowledge, I believe it is vastly less probable, on our present evidence, than is its denial" (Plantinga 1998, 685). Likewise former atheist Anthony Flew finds "improbable" the natural origin of life. "I have been persuaded that it is simply out of the question that the first living matter evolved out of dead matter and then developed into an extraordinarily complicated creature" (Wavell and Iredale 2004).

Scientists working on the topic – and there are many who do work on it – would beg to differ. Nobody pretends that we now know the precise ways in which life formed naturally, and indeed at times the problems seem to grow with every day the subject is studied. But progress is being made and the general opinion is that, given we have today the theories and tools of molecular biology enabling us to penetrate right down into the ultimate workings of the living organism, now is precisely not the time to concede defeat and go home. It is certainly not the time to make room for miracles or other non-natural events. In the years since Darwin, probably the most important realization is more philosophical than scientific: it is about the nature of the problem. When we speak of things as living, what precisely do we mean? What is the difference between something living and something not living?

The obvious answer, and this goes back to Aristotle, is that living things have a kind of substance not possessed by the non-living, some kind of life force. Interestingly, after Darwin, this idea gained a major new lease on life, thanks particularly to its promotion by the German embryologist Hans Driesch, who called the force the "entelechy," and the French philosopher Henri Bergson, who called the force the "élan vital" (Ruse 2006a). But attractive as this thinking was to many, it went nowhere because it really was not very useful. It is all very well speaking of a life force, but unless you can identify it and show what it is doing, why bother to take it seriously? There may be little green men standing in the middle of my living room, but if they are invisible and untouchable and un-whatever-elseable, who cares?

A more fruitful approach has been to think of life not as a something but as an action, a way of doing things, because living things are organized in different ways from non-living things. The English biologist J. B. S. Haldane, working in the first half of the twentieth century, acknowledged that content matters: "It is important to know this, as it is important to know that life consists of chemical processes. But the arrangement of the words is even more important than the words themselves. And in the same way, life is a pattern of chemical processes." Not any kind of process, obviously, but processes that are able to keep themselves going and to replicate their organization. "This pattern has special properties. It begets a similar pattern, as a flame does, but it regulates itself in a way that a flame does not except to a slight extent. And, of course, it has many other peculiarities. So when we have said that life is a pattern of chemical processes, we have said something true and important" (Haldane 1949, 62). In other words, the key to life is not so much substance, but the way in which elements are put together. (Incidentally, Haldane, like A. I. Oparin, the other major figure in kick-starting today's origin of life studies, was a Marxist. But while it is true that Marxism stresses organization, so do other philosophies, and in any case, both Haldane and Oparin started speculating about life before they became full Marxists.)

So, where do we stand today on origin of life studies? Very well known is an experiment that occurred in the early 1950s in Chicago (Miller 1953). By putting inorganic materials in situations that were thought to simulate early conditions on planet earth, and then subjecting them to electrical sparks, as if lightning were striking, two researchers (Stanley Miller and Harold Urey) were able to produce some of the key building blocks of life – molecules known as amino acids, that go to make up proteins, which themselves are both the materials of which cells are made and the catalysts that drive cell processes. Unfortunately, exciting and provocative though this experiment was, it was not followed by similar work tracing through the manufacture of more complex cell parts, ultimately to functioning life. Unsurprisingly, the big problem is that of the chicken and egg: once you have the parts you can create life, but without life how do you get the parts?

Given their key role in heredity, much attention has been paid to the nucleic acids in the cell. Recently, attention has shifted from the DNA molecule to the RNA molecule (Szostak, Bartel, and Luisi 2001). It is this that takes the information from the DNA molecule and then acts to gather up amino acids and build the long chains that form proteins. What has put the RNA molecule in the spotlight is that, in some organisms, it carries the genetic information without needing a DNA molecule. Experiments like those of Miller and Urey have produced not only amino acids, but also the building blocks of nucleic acids. So, if there were some way of linking up these blocks to make RNA molecules, the whole life process might get going. And in fact there are ways to link up parts to make RNA molecules – ways that might well occur naturally. In particular, certain inorganic clays attract the nucleic acid parts and on the surface of such substances they can link up through non-directed processes.

There are still an awful lot of gaps in this account, first off the fact that some people now doubt whether life did (as Darwin assumed and as the Miller-Urey experiments simulated) start in the open, in ponds exposed to the atmosphere. Some now suspect that the deep-sea vents in the oceans, where hot minerals erupt from under the seabed, might be a better bet for life's beginnings (Wachtershauser 1992). However, such suggestions are not science stoppers. They are, rather, issues for debate and discussion, as are such questions as how the inner cell developed a protective shell. The important thing is that there is much work that can, must, and is being done. Particular attention is being paid to the ways in which the RNA molecule can, as it were, improve itself. Some classic experiments show that such molecules can compete against each other, and that better ones (in the sense that they are better able to function and reproduce) are selected. The further you go down this line, the closer you are to a successfully functioning life entity (or proto-life entity).

Even naturalists like Popper (1974) doubted that we would ever crack the mystery of life. "An impenetrable barrier to science and a residue to all attempts to reduce biology to chemistry and physics." No one has yet shown him wrong, but there are those today – perhaps with hopes of Nobel prizes in their minds – who are working flat out to show that this is a barrier that can be penetrated.

Now for the first of Darwin's topics, namely instinct. More generally, in the light of today's interests, we start with the whole area of which instinct is but a part, and turn to the study, from an evolutionary perspective, of animal social behavior, a field that was given its name in a major survey – *Sociobiology: The New Synthesis* (1975) – by the Harvard student of the social insects Edward O. Wilson. Remember how Darwin solved the problem of hymenopteran sterility (the worker ants, bees, and wasps) by arguing that natural selection can act on the family as a unit. This issue was an aspect of a broader problem that worried Darwin for many years, and about which he never truly felt satisfied (Ruse 1980). Go back to natural selection and to the struggle which brings it on. Who is struggling with whom and who therefore gets selected? Leave, for a moment, the talk about genes – this will soon prove pertinent – and go back to things as seen in the *Origin*. Darwin acknowledged fully that sometimes the individual is struggling with brute nature, but often one individual is struggling against another. Clearly when you have a predator chasing a prey, a wolf after a deer, there is no big conceptual problem. But what about organisms of the same species? Are they always in competition with each other, one against the rest? Sometimes organisms in the same species do compete. That is the whole point of sexual selection, which is always about intra-specific struggle. But must there always be struggle within the group, or can group members sometimes help each other or even work together against the members of other groups? Is selection always focused on the individual as the ultimate unit, or can it sometimes be focused on the group? (Today these are known as "individual" and "group" selection respectively.)

Darwin always inclined strongly to an individual perspective. Whether this can be traced to his background in the Industrial Revolution – dog eat dog – is a matter of speculation, but the fact is that he was by inclination an individual selectionist. It certainly comes through in the *Origin*, and is one of the major reasons why he did not want to argue that hybrids are sterile for the sake of the parent species. It may be better for the species not to have hybrids competing, but it is certainly not better for the individual parents. He argued this matter at some length in the 1860s with Wallace, who always favored group-selective explanations. (Again, background might be significant: Wallace was a lifelong socialist.) Nevertheless, Darwin sometimes felt compelled to edge towards group-type explanations, one of which he advanced in the context of humans (we will discuss this in a later chapter). Another instance was the case we have seen where he broadened the unit from the individual to the family. This is not really such a stretch. Even the most hard-line individualist admits that a parent caring for a child can be showing behavior promoted by natural selection. Those birds that fake having a broken wing to draw away the predator from their offspring are obviously promoting their own reproductive chances: no healthy, fertile offspring, no being the fitter or (as is said) showing fitness.

This is not to say that Darwin had conceptual tools that allowed him to go right into the hymenopteran nest and show how the sterility works, or why (for example) there are sterile worker females but never sterile worker males. This all had to wait a hundred years, until in the early 1960s the British graduate student William Hamilton (1964a, 1964b) found the answer. Hamilton pointed out that the Hymenoptera are an exception to the usual Mendelian situation that we were discussing in the last chapter. Normally, individuals have both mothers and fathers, one half set of chromosomes from each. In the Hymenoptera this holds of females, who have both mothers and fathers and hence have a full set of genes, one set from each parent, but not for males, who are born of unfertilized eggs, and hence have only a half set of genes, entirely from mother: they have no fathers. This has the odd result of making females more closely related to sisters than to daughters.

Look at the accompanying diagram (Figure 5.1) and you can see that mothers and daughters have a 50 percent relationship, which is the norm and is found in organisms like humans, whereas sisters have a 75 percent relationship. To express this in the language of the genes, mothers and daughters share 50 percent of their genes, and sisters share 75 percent of their genes. Since in modern terms being a selective success – being fitter than others – means passing on more copies of one's own genes than others pass on copies of theirs, Hamilton pointed out that in fact hymenopteran females are better off raising fertile sisters than fertile daughters. In other words, natural selection does not have to work at the group level. At the individual level, selection will favor sterility because it means that as a hymenopteran female you will be doing better in the sister-raising job than in the

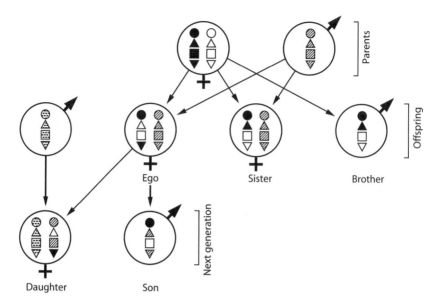

Figure 5.1 Relationships in Hymenoptera

daughter-raising job. No such distorted ratios exist for males and so, as expected, there are no sterile worker males.

Hamilton's work is a triumph, not only of evolutionary thinking, but also of the use of natural selection. Since his initial path-breaking work, a good deal of work has been done to show how social behavior is explicable in Darwinian (neo-Darwinian, if you insist) terms – that is, terms that stress advantage to the individual rather than the group. Much of this has involved the use of game theory, that branch of mathematics that deals with strategies to take against opponents who are competing for shared or limited resources (Dawkins 1976; Maynard Smith 1982). Often, here, model building is of a rather different kind from that which one finds in population genetics. No one knows the genotype nature and so the modeling is of phenotypes instead. This does not mean that population genetics is now theoretically irrelevant, or that sociobiology is not really part of the evolutionary synthesis. No one denies that ultimately it is the genes at work. Rather, at the level of observation at which one is working, it is the phenotype that is known and studied, with the genotypes presupposed and perhaps to be considered later.

The Consilience: One

	Average payoff against Hawk	Average payoff against Dove	Overall average payoff
HAWK	1/3(–100)	2/3(100)	33.3 = 1/3(–100) + 2/3(100)
DOVE	1/3(0)	2/3(50)	33.3 = 1/3(0) + 2/3(50)

Figure 5.2 Matrix of hawks versus doves in an evolutionarily stable strategy: the matrix assumes a starting point of one-third hawks and two-thirds doves, that a victory is worth 100 points but a lost fight costs 200 points, and that non-violent competitors split the spoils

Particularly important for this kind of work has been the idea of an Evolutionarily Stable Strategy (ESS). Selection deploys this strategy where different forms in a population are competing against each other to ensure that the ratios of the forms to each other stays constant – in other words, no form has a selective advantage over the others. Suppose we have two kinds of birds, doves and hawks. Doves fly away when threatened, hawks fight. (In real life, apparently, doves are quite aggressive.) By attaching simple numbers to the various possible scenarios, you can show that at a certain ratio the proportion of doves to hawks will remain stable (Figure 5.2). Selection will promote equilibrium, just as it does in the case of balanced heterozygote fitness. There is no selective advantage to changing from one form to the other. Incidentally, the mathematics is the same if you assume a certain proportion of doves to hawks, or if you assume that each individual shows a proportion of dove-type behavior and a proportion of hawk-type behavior.

Notice that this really is an instance of individual selection over group selection. Sex ratios are another instance of an ESS. In a group, if the number of males is far fewer than the number of females, then males will be at a reproductive advantage and it pays to have mutations that make males more common – until there is a balance. But from the point of view of the group, this is not necessarily a good thing. Males tend to do little or nothing when it comes to child rearing. From the group perspective you might well be better off with 90 percent of hard-working females and just 10 percent male studs, who can do all of the fertilizing quite efficiently. This raises a question about group selection in itself. Is it always wrong to take a group-selective perspective and if not when is it legitimate? I doubt that any evolutionist would want to say absolutely that group selection is always impossible (Reeve and Keller 1999). It is just a question of the circumstances. For instance, if a population tends to fragment into small groups and there is a real benefit from sacrificing for this group, then in the short run group selection can probably work. And if the groups are short-lived and constantly re-forming, then it would seem possible for a kind of group selection to be important overall. There has been much discussion among Darwinians over the origin and maintenance of sex (Ruse 1996). It has advantages from a group perspective – new, favorable mutations can be spread quickly in the population. However, from the individual viewpoint of females it seems disastrous because, as noted above, they tend to do the work. Why propagate someone else's genes if he makes no contribution to child rearing? Some Darwinians argue that there must be an individual-selection reason for sex. Hamilton argued that it led to the shaking up of genotypes and that this gave immunity against parasites, which generally evolve very much faster than their hosts (Hamilton, Axelrod, and Tanese 1990). John Maynard Smith (1978), on the other hand, was convinced that group benefits must be involved in keeping sex going. He could not see that individual selection could do the job.

But to ask a question as much philosophical as scientific, should one automatically prefer an individual-selection explanation over a group-selection explanation? Opinion divides. Some argue that one should (Williams 1966): they contend that individual selection explanations are "simpler" and are part of the laudable, reductive trend in science since the sixteenth and seventeenth centuries. Note that here we are using the term "reduction" in a sense slightly different from before. We are not talking about deducing something from something else: group selection is not be deduced from individual selection – if anything it is being pushed aside. The meaning here is, rather, that in a scientific explanation the smaller is preferred over the larger – molecules over gases, atoms over molecules, genotypes over phenotypes. The former, lower levels are thought more powerful in some way than the latter, higher levels (for smaller and larger entities respectively).

The Consilience: One

Not everyone, however, thinks that these sorts of considerations are definitive. Simplicity is a notoriously slippery customer. Is it just something in the eye of the beholder, something that we like because our puny minds are not capable of grasping the real complexities of nature? And why smaller over larger – at least, why, in principle, smaller over larger? Marxists and others of a similar bent argue that the world consists of layered complexity, and there can be no reduction, in the sense just mentioned, of the larger to the smaller (Engels 1964). Higher levels of complexity have their own meanings and connections, closed to the lower levels. The same could well be true of the individual/group selection debate. Perhaps some things just are susceptible only to a group explanation (Sober and Wilson 1997). In any case, individual-selection explanations have their own distinctive problems. Perhaps the most popular and powerful metaphor in modern biology, introduced by the English sociobiologist and popular writer Richard Dawkins (1976), is that of the "selfish gene." What Dawkins meant to do was to draw attention to the way in which selection promotes the individual over the group. But very readily one starts switching from selfish genes to selfish people. Before long one starts to give content to the suspicion that Darwin and his followers were reflecting less the facts of nature and more their own commitment to a harsh, Victorian, laissez-faire, socioeconomic viewpoint. Everything that we think and do is a matter of self-interest, and since it is natural there is nothing we can do about it.

We are not going to end this discussion here. The individual selectionists have a good point. Selfish genes do not necessarily equal selfish people. A mother bird may be acting in her own genetic interests in pretending to be injured in front of a predator and thus drawing attention from her nest, but by no stretch of the imagination could she be called selfish. You might object that, being just a bird, she should not be called anything moral, but, wherever you stand on this, "selfish" is not an appropriate term. The group selectionists have a good point. Too often people mix up their levels and start talking about selfish organisms when they should be talking about selfish genes. Metaphors are powerful things and they do tend to have lives of their own. It is simply not true that selfish gene talk has never made anyone think this applies at upper levels. To quote one sociobiologist: "Scratch an altruist and watch a hypocrite bleed" (Ghiselin 1974a).

At the moment in the scientific world the individual selectionists have much the upper hand. The simple fact is that individual-selection theory has paid massive dividends when it comes to under-standing animal social behavior, and no one yet sees an end to this. It is not a question of philosophy. Individual selection works in a way that group selection does not. Having said this, we must remember that these are in major part empirical questions, and the tide could change. Already Wilson, who at first endorsed Hamiltonian kin selec-tion with enthusiasm, has argued that in the more sophisticated social insects it is no longer appropriate to use individual selection, that, at the least, one must take Darwin's position and consider the nest as one "superorganism," and that perhaps a group approach may be demanded (Holldöbler and Wilson 1990).

Paleontology

The fossil record, since the Cambrian, was sketched out before the *Origin of Species* appeared on the scene (Bowler 1976). After 1859, paleontologists very quickly took up an evolutionary interpretation, agreeing with Darwin that the record is the history of life as it devel-oped through the ages, and that the gaps do not reflect times when life was interrupted but rather times when the record was not laid down. Not that people were particularly Darwinian with respect to causes, and much of the theoretical work was marred by the simplistic analogies we have already referred to between embryology (ontogeny) and paleontology (phylogeny). People were more inter-ested in working out paths of evolution – phylogenies – than causal processes (Bowler 1996). But the lack of deep thought was sheltered and in the end brushed aside by those massive new discoveries that were being made, especially in the American West, and so pale-ontology grew and developed, becoming in the popular mind the paradigm of an evolutionary subject.

What is really puzzling is that the acceptance of evolution made so little difference to the actual practice of paleontology. Rarely if ever in the fossil record did one get a smooth transition from one form to another. It was almost always a jump, big or small, from one form to another. Hence, people dug out fossils and tried to see which forms were most alike and which came together in the record. People compared fossil forms with modern forms. People looked at the embryos of modern forms and sought analogies with the forms of long ago. Methodologically, nothing much changed. A non-evolutionary version of the biogenetic law was almost an axiom before the *Origin* and evolution. We have heard Louis Agassiz, the great opponent of evolution, on this subject, with his threefold parallel: the history of life, the history of the individual, and the range of living things in the world today. If you hold to this, whether or not you believe there are actually links really makes little difference to your daily work as a scientist. As a result, one often cannot tell with the post-*Origin* paleontologists which paper marks the transition from non-evolutionists to evolutionists. Imagine if one could not tell when a geneticist had read the paper on the double helix by Watson and Crick!

What is going on here? It really looks less like a scientific shift and more like a metaphysical one. People were turning from a worldview that allowed interventions by the Creator – in fact, demanded such interventions as forms changed from one to another – to a worldview that refused to allow the Creator any direct role – in other words, a naturalistic world picture. Gaps in the fossil record were explained first as genuine, and reflective of the Creator's actions, and then as the product of incomplete fossilization as organisms evolved. But really, one might think, this was essentially a metaphysical shift, from supernaturalism to naturalism.

There is certainly some truth to this way of looking at the coming of evolution, and it was not alien to the scientists of Darwin's day. Indeed some, the naturalists certainly, welcomed evolution precisely because they endorsed the metaphysical shift they thought it embodied. Thomas Henry Huxley (1858) particularly was scathing about his biological rival, Richard Owen, and accused him of thinking in terms that were no longer acceptable for a scientist. Huxley and others saw themselves as making a conscious move to naturalism, and evolution was part and parcel of this. But there is a little more than just metaphysics at stake, if by this one means simply faith without reason or justification.

On the one hand, naturalism had worked very well in the physical sciences. It was not stupid or irrational to want to apply this to the biological sciences. And if one did so, with evolution rolled in, then one could go a long way: for example, it became possible to explain things like biogeography without appealing to miracles. On the other hand, taking up evolution was not just a matter of naturalism. Evolution could be empirically false on a naturalistic worldview. It was not logically impossible that organisms sprang in full form from dust, or that the world is eternal, or that it was seeded from outer space, or some such thing. None of these are likely, and Pasteur for one was making some of them increasingly unlikely. But Pasteur did not simply theorize, he proved his point empirically. Perhaps you might want to say that evolution benefited from the fact that, long before Darwin, people had started to exclude extraterrestrials and the like, but this does not make the exclusion any less empirical. So, although the two were certainly linked, it is hardly fair to say that the acceptance of evolution was the inevitable consequence of a move to naturalism.

And even if the move to naturalism was a major factor in Darwin's support, time did not stand still. As the nineteenth century ended and the twentieth century ground into action, unreliable techniques were discarded and new methods of ferreting out the past were adopted. Slowly and steadily, the history of the past was unfurled, and indeed it goes on unfurling to this day (Knoll 2003). Now, thanks to the physical sciences, it truly is possible to give absolute dates to events in the past. The universe is at least 15 billion years old and the earth is about 4.5 billion years old – a far cry from those limited spans that troubled evolutionists in the years after the *Origin*. More than this, major gaps in the record have been filled. Most particularly, the time that so worried Darwin – the Precambrian era – need worry us no more. There is evidence of life going back beyond 3.5 billion years, until indeed the very earliest time that one might expect life to appear. From then on, one sees evidence of life's development up until the time of the Cambrian, rather more than 500 million years ago (Figure 5.3).

Obviously much of this evidence is indirect and has to be inferred, but it is there nevertheless. For instance, one of the chief divisions in living things is between simple, single-celled organisms, prokaryotes, which have no nucleus and the ribonucleic acid loose in the cell, and eukaryotes, single or multicelled organisms, which have cells with nuclei and chromosomes and so forth – the kinds of cells we considered in the last chapter. It is thought that eukaryotes evolved from pro-karyotes: Lynn Margulis (1970) proposed a brilliant hypothesis about

Life on earth	Time (billions of years)	History of our planet		
End of dinosaurs, age of mammals	0.065			
	0.3	Pangea (300mya–180mya)	(Plate tectonics)	Free oxygen in atmosphere
First well-mineralized skeletons	0.57			
First body fossils	0.70			
		Cranial sediment chiefly oxidized		
Possible early eukaryotes	2.10		(Plate tectonics probable)	
		Most banded iron formations		
Prokaryotes becoming diverse	2.21			
		Cratonal sediment chiefly unoxidized		Chiefly anoxic atmosphere, probably reducing
Oldest microfossils	3.80	Rocks chiefly granitic and gneissic, sediments extensive	(Global tectonics unlike present)	
	4.4	Oldest dated mineral (zircon)		
	4.56	Origin of earth		

Figure 5.3 The age of earth

how some prokaryotes absorbed other prokaryotes and thus a more complex cell was formed. A major difference between prokaryotes and eukaryotes is that the former get their energy by methods like (and including) fermentation, by breaking down glucose, whereas the latter get it by respiring, that is by burning glucose in oxygen. The two processes are not entirely different, and significantly the latter seems to be built on the former. However, you do need oxygen for respiration. As best we know, there was none in the early earth atmosphere.

So presumably it was created, and presumably this means (as today) photosynthesis. In other words, the eukaryotes both pose the puzzle and solve it. You need oxygen to respire; some eukaryotes can perform photosynthesis and thus produce oxygen. The cyanobacteria, the blue-green algae, can also perform photosynthesis, and they seem to have been around from before the eukaryotes arrived but after the prokaryotes arrived. Adding to the hypothesis, the algae have metabolisms half way between fermentation and respiration.

In absolute dates the cyanobacteria leave fossil traces around 2.7 billion years ago (bya); the first unambiguous eukaryotes leave fossil traces around 1.2 bya, by which time they were on the way to being multicellular. This means one would expect to see that the cyanobacteria had pumped up the oxygen enough for the eukaryotes around 2 bya. And, as expected, one finds geological evidence that before this time the rocks were in an oxygen-free environment and after in an oxygen-filled environment. Iron without oxygen does not rust; iron with oxygen does rust.

> Some of the most compelling evidence for oxygen scarcity on the early Earth comes from gravel and sand deposited by ancient rivers as they meandered across Archean and earliest Proterozoic coastal plains. Pyrite [FeS_2 – fool's gold] is common in organic-rich sediments, forming below the surface where H_2S produced by sulfate-reducing bacteria reacts with iron dissolved in oxygen-depleted groundwaters . . .
>
> The same is true of two other oxygen-sensitive minerals: siderite (iron carbonate, or $FeCO_3$) and uraninite (uranium dioxide, or UO_2). Neither of these minerals is found today among the eroded grains that make up sediments on coastal floodplains, but both occur with pyrite grains in river deposits older than about 2.2 billion years.
>
> (Knoll 2003, 97)

But after 2.2 bya, do we get the deposition of minerals that can form only in the presence of iron? Yes, we most certainly do! A nice example is that of the red sandstones of the Grand Canyon. "These rocks – called red beds, in the button-down parlance of geologists – derive their color from tiny flecks of iron oxide that coat sand grains. The iron oxides form within surface sands, but only when the groundwaters that wash them contain oxygen. Red beds are common only in sedimentary successions deposited after about 2.2 billion years ago" (Knoll 2003, 97). Before 2.2 bya, the oxygen level cannot have

The Consilience: One

Figure 5.4 Skeletons of Archaeopteryx (*right*) and pigeon (*left*)

been more than about 1 percent of today's levels; after 2.2 bya, levels were at least 15 percent of those we find today. Just what one would expect if this was the point at which the eukaryotes really got going.

Today, over and above any connection between reading the fossil record in an evolutionary fashion and a commitment to naturalism, the record itself, from the post-Cambrian on, has filled up in ways that strengthen the link empirically. Impressively, there are many, many examples of links between major groups of organisms. Most recently, in the north of Canada, researchers have unearthed a vertebrate that seems half way between sea-dwelling fish and land-dwelling (at least land-using) amphibian (Wilford 2006). Going back in time, the most famous link is Archaeopteryx, the little beast that is half reptile and half bird – a small dinosaur with feathers (Feduccia 1996) (Figure 5.4). In fact, the first feather was discovered even before the *Origin*, but until the whole animal was found no one realized its significance. There are also many examples of chains of fossils, showing the change from one form to another. Huxley (1877) made much of the incredible findings of horse evolution in America: having a chain from a four-toed ancestor to the present one-toed horse, he predicted that further back in the record there would even be a five-toed horse. It was very exciting when, shortly after this prediction, just such an animal

(Eohippus) was discovered. Recently, and satisfyingly, the move from land animals to the sea mammals have been uncovered in great detail (Carroll 1997). In a later chapter we shall turn to the human fossil record which is now being revealed in ways that would never have been imagined in Darwin's day.

Paleontology has many triumphs. Yet there is a paradox. Virtually every layperson, if asked why they believed in evolution, would answer that the fossil record – all of those dinosaurs – was the main reason. Yet virtually no professional evolutionist outside the field would think first of paleontology if they were asked about belief in their subject. Why is there such a distance between the popular and professional perceptions? More precisely, why would other evolutionists not rate paleontology as would outsiders? Why do pale-ontologists resent this and often give their professional colleagues reason to feel irritated? It is basically the question of causes (Stebbins and Ayala 1981). If you are a fruitfly geneticist, you work with the genes and natural selection. You take this as your basis – your null hypothesis – and move outward from there. In paleontology, the needs for genes and selection are apparently a lot less pressing. For a start, you have all of the work of digging out the fossils and making sense of them. And in the end, if you are not bothered about selection, you can always go for something else. The molecules are long gone, so no one is going to do any genetic fingerprinting on Stegosaurus. Take the titanotheres, lumbering extinct mammals from North America with baroque appendages on their noses (Figure 5.5). Why bother with selection? Why not go for some kind of momentum theory, which says that organisms reach a peak and then go into a decline as they get overly complicated?

No wonder non-paleontological professionals get irritated, and why paleontologists can seem so obtuse to their fellow evolutionists – chip-on-the-shoulder obtuse. A paleontologist really does not have a say when it comes to causes. It is the fruitfly-genetics type of evo-lutionist who proposes the causes – rather than Indiana Jones it is the nerd from the Gary Larson cartoon. At best, paleontologists are left to fit their findings in with what others say. They never get to make up causes in their own right, or if they do they are laughed at by the fruitfly types. At the time of the making of the synthetic theory of evolution, paleontology was apparently brought firmly into the

Figure 5.5 Titanothere

Darwinian family. More than anyone, the American mammalogist G. G. Simpson (1944) argued and showed how population genetics can and does lead to the kinds of patterns that we find in the fossil record. Obviously, paleontologists could still not get their hands on the ultimate units of heredity, but thanks to Darwinian ideas they could start to make sense of many things. For instance, the protuberances on the faces of titanotheres call out for an explanation in terms of sexual selection. You may not get turned on by them, but then you are not a female titanothere.

And, more broadly, when the paleontologist turns to phenomena that are only going to be revealed by looking at changes over long periods of time, Darwinism can offer much by way of explanation. For instance, the American paleontologist J. John Sepkoski (1978, 1979, 1984) showed how the patterns of life seem to follow sigmoidal curves – great explosions of life followed by flattening off and even long decline, followed in turn by the explosions of new forms repeating the pattern (Figure 5.6). It is easy to see how this almost demands a selective explanation. New niches are opened by new adaptations and are exploited to the full; then, as they get crowded, the rate of innovation and new-form creation declines and remains at best

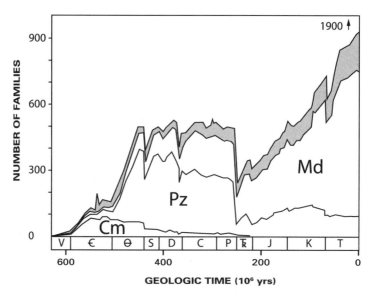

Figure 5.6 The Phanerozoic history of the taxonomic diversity of marine animal families. The upper curve shows the total number of fossil families known from direct evidence or range extension to occur in each stratigraphic stage of the Phanerozoic. The number "1900" in the upper right-hand corner is the approximate number of animal families described from the modern oceans; these include a large portion of soft-bodied and lightly sclerotized groups, such as medusoid coelenterates and many small arthropods, which are only rarely preserved as fossils. The fossil diversity of these "poorly preserved" families (which largely reflects range extension between various exceptional fossil deposits and the Recent) is indicated by the shaded field in the figure. The two curves below the shaded field divide the diversity of heavily skeletonized families into three fields, representing the three "evolutionary faunas" that dominate total diversity during successive intervals of the Phanerozoic: the Cambrian fauna ("Cm"), the Paleozoic fauna ("Pz"), and the Mesozoic–Cenozoic, or Modern, fauna ("Md"). Symbols along the horizontal axis indicate geologic systems (for example, "V" = Vendian). (Sepkoski 1984, p. 249)

stable, especially as a new breakthrough occurs. Sepkoski argued that this is just the sort of thing that happened in the Cambrian, as new forms were developed and successors then multiplied to fill the ecological spaces that had just been opened up.

Punctuated equilibrium

Yet, as Tolstoy taught us, not all families are happy all of the time. The Darwinian family was made unhappy by paleontology. No sooner had this problem child come in from the cold and rejoined the family, than it once again showed its discontent and lack of appreciation for the achievements of its siblings. Not to mince words, the past fifty years have seen cross talk, misunderstanding, and even outright hostility between paleontologists and other evolutionists (Ruse 2000; Sepkoski and Ruse 2008). The chip-on-the-shoulder resentment is still there – with some justification. Paleontologists point to facts that simply cannot be established from micro-studies, and that surely should be considered if we are into macro-studies of the full sweep of evolution. For instance, we are now aware of many evolution-affecting events unconnected with the organic processes themselves. From the point of view of selection, they are random if not uncaused. Most importantly, we know that life's history is dotted with times of massive extinction, which seem to have little direct biological cause. The most famous was that which saw the end of the dinosaurs about 65 million years ago. There is much debate about the causes of these extinctions. This particular one was undoubtedly caused by a comet or asteroid falling to the earth just off Mexico and causing dreadful upheavals. Earlier extinctions may have had similar causes or may have been the side-effects of terrestrial geological phenomena – more precisely, the shifting of the continental plates (to be discussed in more detail in the next chapter), which in turn brought on dramatic changes in habitat.

Paleontologists rightly argue that other evolutionists must think about these sorts of phenomena, or respect those who do. Earth-history events are an essential part in the completion of the story of life on earth. Perhaps a less justified aspect of the paleontological attitude is the continued envy of those who work at the cutting edge of causal studies. Students of the fossil record have trouble coming to terms with the fact that this is not truly their field. Instructive in this respect is the so-called theory of "punctuated equilibrium," championed by the two paleontologists Niles Eldredge, American Museum of Natural History curator, and the Harvard professor Stephen Jay Gould, until his death Richard Dawkins's only equal at explaining evolutionary ideas to the general public (Eldredge and Gould 1972). Natural selection implies that change will be smooth and gradual. If you have major jumps – saltations – more often than not they are disastrous for their possessors and are wiped out in the first generation. The key to evolution is: one little bit at a time. Gaps in the fossil record, therefore, must be the result of missing links rather than indicative of real jumps or leaps. The crux of the Eldredge–Gould claim – their notorious claim – was that the fossil record is a lot more complete than most Darwinians suppose. Hence its jerky and rapid-change character reflects what really happened. In other words, traditional thinking stands in need of significant revision.

But from what to what? Much of the trouble here was the great ambiguity about what precisely was being claimed and what was being demanded. If the claim is that the fossil record tends to go by fits and starts, then it is not entirely obvious who would ever disagree with this. No one, certainly no Darwinian, thinks that selection pressures will always be entirely uniform or denies that evolution will progress at different rates at different times. If the claim is more drastic – namely, that never are there linking fossils between the forms we have – then it would almost surely be that case that natural selection needs supplementing in some fashion. As it happens, few except extremists would argue that there are never linking forms. The question therefore becomes why there are so often no linking forms. Suppose you accept, with Eldredge and Gould, that it is not enough simply to write things off as a function of the inadequacy of the fossil record, the question that still must be answered is: What kind of supplement is needed to the theory of natural selection? Is it an extension of selection, something different but not threatening to selection, or something that simply shows that selection is inoperative at crucial times in evolution's history?

Gould, particularly, often floated the suggestion that something new – perhaps selection-denying, or at least selection-replacing – was needed. He even went so far as to say: "The synthetic theory of evolution is dead" (Gould 1980). In other words, he was staking a claim that he, as a paleontologist, had the need and the right to make a causal contribution to the full picture. Unfortunately, for all of this bold talk, Gould was never really able to say what his contribution might be. For a while, he floated the idea of saltations, but then, when criticized by the geneticists, retreated and claimed that he was being misunderstood. There was a lot of heat but not a great deal of light.

At a more moderate level, many evolutionists think the jerky record requires perhaps modification and augmentation of traditional selectionist thought, rather than rejection.

In fact, there was already pertinent material waiting to be used – material with which Eldredge and Gould toyed in their earliest writings, but then apparently abandoned as insufficiently radical. There has long been considerable interest in the reasons for the creation of new forms – new species – and much has been written in the Darwinian tradition, trying to update it in the light of modern genetics. Ernst Mayr (1954) proposed what he called the "founder principle." There is always a great deal of variation in populations. Newly isolated small groups will therefore almost necessarily have only small subsets of the possible variations of the whole group. As these founders interbreed, on the one hand they will often experience strong new selective pressures in their new surroundings, and on the other hand their genes will have to settle down and work together. This in itself is likely to promote rapid change, as a form of genetic revolution takes place, until things have stabilized. Hence, for these reasons – some obviously directly Darwinian, others more a factor of chance (although certainly chance of a kind that Darwin would have recognized and accepted) – change from one form to another will often be very rapid and not likely to be reported in the fossil record.

Especially if this is the usual way in which new forms are created, then the jerky fossil record is almost what is expected. And so, having come full circle from consistency to opposition and back to consistency, let us leave things at that. Or, rather, let us use this as the springboard for future discussion. No paleontologist would deny the relevance of Darwinism. No paleontologist would deny the great relevance of Darwinism. Some paleontologists (and fellow evolutionists) would argue that paleontology fits comfortably within the Darwinian picture. Others feel that more by way of causal explanation is needed. How much of this yearning is a function of real issues and how much of discipline insecurities, we will have to leave to history to judge.

6

The Consilience: Two

We move now through the other topics that Darwin considered in this part of the *Origin*. Overall, I do not think there was any absolute reason why Darwin ordered the discussion as he did. All of the parts seem to bear the same relationship to the core of evolution through natural selection, although, for obvious reasons, it made more sense for Darwin to discuss morphology before embryology, because the latter presupposes understanding of the former.

Biogeography

Biogeography looks at the distributions of organisms across the globe and seeks the underlying causes. It combines ecology – the short-term movements and interactions of organisms today and in the past – and evolution – the heritable changes that occur to organisms in the long term. Most researchers do not look on these differences as marking any significant boundaries – ecology in the long run tends to turn into evolution – and people move smoothly from one area to the other. The evolutionist Edward O. Wilson, the world's leading authority on the social insects, is also one of the authors of a very important ecological theory about the equilibrium of species on oceanic islands (MacArthur and Wilson 1967). In seeing no great differences between the fields in which he works, Wilson (working with theoretical biologist Robert MacArthur) was following Darwin who, in the *Origin*, likewise comfortably moved from evolution to ecology and back again. Remember how Darwin was particularly interested in the ecological relationships between organisms in the struggle for life, and how he gave the example of clover plants, destroyed because the bees that pollinate them are preyed upon by field mice, which are in turn kept in check by domestic cats. Wilson's equilibrium theory, showing the intimate relationship between ecology and evolutionary theory, was picked up by his student J. John Sepkoski Jr., who applied it to the occupation by organisms of empty niches, through time as well as through space, and who was then able to produce and explain the equilibrium patterns we saw in the last chapter.

Biogeography was one of the areas of greatest support for Darwin in his own time and it continues to have that role today. We have already examined the sickle-cell anemia distribution and seen why it maps the instances of the disease malaria. There are many, many other similar cases. One of the most famous in the history of evolutionary theory is that made much of by Ernst Mayr in his classic *Systematics and the Origin of Species*, first published in 1942. Mayr, who had just spent ten years cataloging bird skins in the American Museum of Natural History and who was thus highly sensitive to minute but significant variations in form, was determined to show (contrary to the saltationist speculations of certain prominent geneticists) that the course of evolution is smooth and gradual, as one would expect if the main cause of evolutionary change is natural selection. Mayr seized on the phenomenon of the "ring of races," in which linked, related populations, encircle the globe (or a wide area), the ends of the ring finally meeting. If one could show that the intermediate adjacent groups within the ring were interbreeding but the end groups were not, then one would have a perfect example of speciation in action. Take out the central populations and you have two separate distinct species, unable to exchange genes. One recently described example is that of the greenish warbler complex (*Phylloscopus trochilodes*) in Asia. It forms a ring of six subspecies (at least, it did until deforestation broke the links), and the molecular evidence confirms not only that neighboring groups are the most closely related but that there is some interbreeding. However, at the ends of the ring, the two groups do not interbreed. They behave like different species, with different songs, and members of the two subspecies simply do not recognize each other (Irwin, Bensch, and Price 2001; Figure 6.1).

Take note of the logic of the argument here. As with the finches and tortoises of the Galapagos, and as with the sickle-cell anemia genes, the rings of races are already in place. No one is setting up experiments

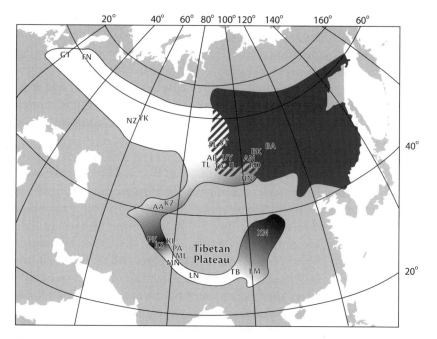

Figure 6.1 Ring of races in the greenish warbler complex. In the striped area the populations overlap but do not breed. The opposite side of the ring has now been broken by deforestation. (Reprinted, by permission, from Irwin, Bensch, and Price 2001)

and then seeing how they run. No one is making predictions about the future. One is taking certain phenomena and then fitting them into place under a single overarching hypothesis about change, bolstered by subsidiary hypotheses – for instance, about the ways that organisms get transported around the globe. Some people seize on this procedure and argue that it shows that the science of evolution is in principle different from other sciences such as physics (Ruse 1973). Often the subtext, sometimes silent, sometimes spoken, is that this means evolutionary biology is somehow inferior. It is condescendingly spoken of as a descriptive or narrative or historical science. A linked argument, one that often occurs in this context, is that evolutionary biology has to be different because (unlike physics and chemistry) it is dealing with unique phenomena. One instance of gravitational attraction is like all others, and one instance of sulphuric acid is like all others. But although Hitler and Napoleon were both dictators, they

The Consilience: Two

were unique individuals and the historian must take this into account. Likewise the evolutionary biologist must take the uniqueness of the sickle-cell gene into account. You cannot think in the same way as one does in the physical sciences.

None of this is convincing. Evolutionary theory is certainly historical, and I suppose it is often descriptive and narrative – although whether one would describe a page of symbols from a population-genetical argument as a fun read is perhaps another matter. The uniqueness argument is based on such an elementary mistake that one suspects nothing would deter its supporters from the conclusion to which they think it points. A microsecond of serious thought shows that everything is unique in one sense, nothing is unique in another. Mars and Venus are unique, but their common features allowed Kepler to bring them beneath law and into science. The Big Bang may well have been a one off, but it is explicable by scientists because it shares features with other natural events. If it did not, you could say nothing about it. Likewise, the sickle-cell gene is unique, but it is not the only case where there is geographical distribution because of disease. There are other blood diseases around the Mediterranean that are likewise connected to beneficial effects in the heterozygotes. (Thalassemia is another form of genetically caused anemia linked to protection from malaria.) The whole point about the sickle-cell case is that, inasmuch as it is brought into science, it is precisely because we can apply the standard theory of population genetics – a theory applicable to all cases where heredity is involved. Sickle cell is unique and it is not unique.

None of this is to deny – as agreed in the discussion of the structure of evolutionary theory – that often we have a somewhat loose argument, certainly nothing deductively tight. But whether this all adds up to an entirely different logic is another matter. In principle, one could have a different logic: indeed, in my discussion of function, which treats sympathetically Kant's claim about the significance of final-cause thinking in biology, I have rather agreed that there is something different about biological thinking. It makes sense in biology to ask: What is the function of the growth on the nose of the titanothere? It does not make sense to ask: What is the function of the rings around Saturn? This does not represent an inherent weakness in biological thinking any more than being a woman rather than a man is inherently weak. The point is that in biology we are dealing with entities – organisms and their adaptations – that demand final-cause thinking. In physics and chemistry, this is not so, except where we have moved into the biological context, in which case end-directed thinking comes at once into play. You may not ask what is the function of the oxygen molecule in water, but you may certainly ask what is the function of the order of the molecules along the DNA chain. What is the genetic code, what is it for, and how do we crack it?

I will return to this issue when we get to religion. For now, we can move on to note that if we are talking about evolutionary theory as such, it is simply not true that it is never experimental or predictive. The alcoholic fruitflies give the lie to this argument. Ayala's group made predictions about what would happen if they upped the alcohol content of the flies' surroundings – namely, that there would be selection for alcohol tolerance (if not use) – and the predictions came out spot on. There are many similar experiments in evolutionary theorizing. Running parallel to this point is the fact that often what we do in scientific testing and confirmation is to retrodict rather than predict. We infer what must have happened rather than what will happen, and then we go back to check. The logic is the same: we are inferring from the premises and then checking. A Popperian will not object, because if the forecast is wrong the premises cannot be right (or something else must be amiss). The difference between the two is just temporal, but retrodiction preserves what makes prediction so important, namely that the best science goes out on a limb rather than simply fitting known facts to a hypothesis. Both go beyond simple explaining after the event, which does make people queasy and rightly worried about the quality of the science.

Thus, for instance, if someone points me toward an offshore island, then I can make all sorts of predictions (retrodictions) or forecasts about what one might expect: that the inhabitants will be a lot more like those on the nearest mainland than on a continent far away; that they will likely show features making them fit for an island – for instance, the insects may have reduced ability to fly (because winds will otherwise blow them away to sea); that if the island is one of a chain, there will be relationships between the inhabitants of the islands that reveal past history. Of course, not every forecast works out, but that is the nature of science. Sometimes, indeed, the forecasts that do not work out are the ones that are really important. If you find that the island inhabitants are not as you expect, then you may be on to something.

The Consilience: Two

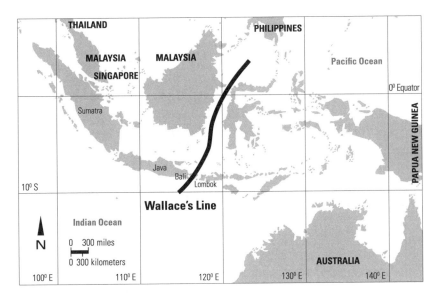

Figure 6.2 Wallace's Line (after Wallace 1876)

Alfred Russel Wallace (1876) discovered that the inhabitants of the East Indies are divided into two separate groups (the division is marked by what is now known as "Wallace's Line"); islands close together on opposite sides of the divide have inhabitants that are quite different; islands on the same side have inhabitants that, though they might be quite separate, are similar. Bali and Lombok are only 20 miles apart, but the birds on Bali are "western" and related to those on Java and Sumatra, and those on Lombok are "eastern" and related to those of Papua New Guinea and Australia. You start to look for reasons – in this case, past geological phenomena as well as ice ages, which reduced sea levels and made transitions possible except across the very deep divides of Wallace's Line (Figure 6.2).

Plate tectonics

There is no great surprise that geology enters into biogeographical explanations. It has always been so. Darwin, following Lyell, saw biogeography as intimately related to geology. A topic of some interest, therefore, is the extent to which biogeography connects with, or reflects,

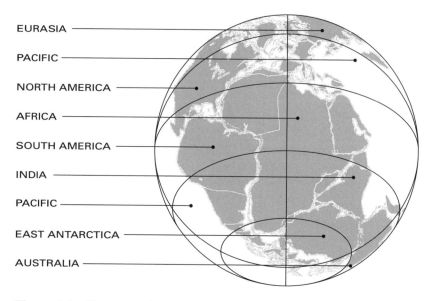

EURASIA
PACIFIC
NORTH AMERICA
AFRICA
SOUTH AMERICA
INDIA
PACIFIC
EAST ANTARCTICA
AUSTRALIA

Figure 6.3 Plate tectonics

the most important event in geology in the twentieth century: the discovery of plate tectonics, pointing to and explaining continental drift. Instead of the lands of the earth staying fixed and stable, they move around the globe on large plates, which are in turn created and ultimately destroyed by the currents in the molten lava beneath the earth's surface. In some parts of the world, plates are being created and pushed up and then along, and in other parts of the world, plates are being sucked or pushed down, and thus destroyed. Continents piggyback on these plates. Hence, the world looked very different millions of years ago from what it does now (Figure 6.3).

The relationship between biogeography and modern geology seems to be two-way. In some respects, biogeography gives support to geology – in much the same logical way that for Lyell and Darwin biogeographical evidence gave support to the theory of climate. At one point in the past, today's continents were joined together in one supercontinent, Pangea. There is lots of evidence for this landmass, not the least of which is biogeographical. One finds distributions of fossils that simply do not make sense unless the continents of today

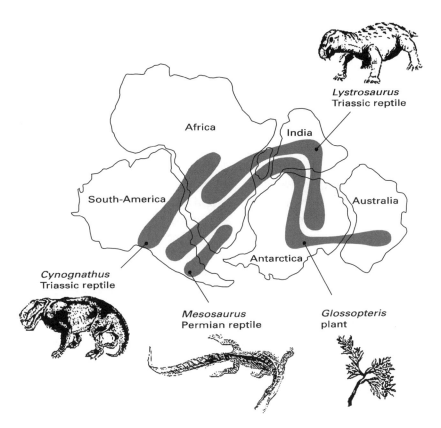

Figure 6.4 Biogeographical distribution (from a U.S. Geological Survey map)

were once joined. To take just one example: look at Figure 6.4. Lystrosaurus, a mammal-like herbivorous reptile, found rather more than 200 million years ago, is fat, short, and squat. It is certainly not a beast that would have been a world traveler. It stayed close to home. Today it is found in the same fossil deposits (Lower Triassic) on the continents of Africa, (Southeast) Asia, and Antarctica. A miracle were it not for the fact that, more than 200 million years ago, all of those continents touched when they were part of Pangea. They have since drifted apart. The fossils (and you will see from the diagram that there are others) support the geological hypothesis.

In some respects, the relationship goes the other way, with modern geology (continental drift) throwing light on problems of evolution.

Consider the southern beeches (*Nothofagus*) to be found in both Chile and New Zealand but nowhere else, or the conifers (*Araucaria*) to be found on both the South American continent and Australasia but nowhere else. Examples like these are precisely what one would expect, given the combination of Darwinian theory and continental drift. A more complex example is that of the interchange of groups between North and South America – nice not just for showing the nature of a biogeographical argument but also because it illustrates yet again that a historical argument is not something of a special kind, at least not because of some mystical quality of uniqueness. As Darwin noted, home-grown North and South American mammals are significantly different. However, in the fossil record there are North American forms in South America – and some of them, or their descendents, are still extant in South America – and conversely some South American forms in North America – with corresponding extant forms (Marshall, Webb, Sepkoski, and Raup 1982). These are not very old fossils: they go back about 10 or 12 million years, with a large jump about 3 million years ago. The reason is now obvious. North and South America were parts of different landmasses, which – thanks to continental drift – moved closer together and about 3 million years ago joined up along the Panamanian land bridge. Before the land bridge formed, when the continents were close together, some mammals moved from island to island across the gap, and then with the join there was a real rush.

At the family level we have about the same number of common forms in both continents (35 in South America and 33 in North America). In time, in both South America and North America, common forms peak (respectively 39 and 35 families) following the immigrations, and then (because of extinction) a decline occurs in overall numbers. "Today, members of 14 North American families occur in South America and contribute 40 percent to the familial diversity of that continent, whereas members of 12 South American families occur in North America and account for a nearly equivalent 36 percent of that continent's familial diversity" (Marshall et al. 1982, 1354). In other words, the interchange seems to be balanced. Although there were increases in the number of families for a while, and although the composition of families was changed very significantly, thanks to extinction, the overall numbers have more or less reverted to the "pre-join" figures, which we know independently were fairly steady. How

Anteaters
Armadillos
Capybaras
Glyptodonts
Monkeys
Opossums
Porcupines
Phorusrhacids
Sloths
Toucans
Toxodonts

Bears
Camels
Cats
Dogs
Elephants
Horses
Peccaries
Rabbits
Raccoons
Skunks
Tapirs
Weasels

Figure 6.5 The great American interchange

does one explain this? Most obviously, one turns to a theory about immigration and extinction and whether or not such changes produce a balance or equilibrium. And here the proper move seems to be toward another (somewhat modified) application of the MacArthur–Wilson island biogeography theory. This theory argues that denuded islands will receive immigrants from the mainland and that after time there will be a balance, an equilibrium, between the incoming members and those that are disappearing through emigration or extinction. "The resultant diversity and the time span required to attain equilibrium are largely dependent on the size of the area; thus continents will have a higher species diversity and lower per-species turnover to attain equilibrium, as compared to oceanic islands" (p. 1355). But we do expect equilibrium and this is what we seem to get, as the theory predicts.

As we start to look at the interchange in more detail, however, the picture gets more complicated. Consider genera rather than families. There is no return to old balance in South America. The number of genera jumps right up. But this seems to be almost entirely a function of the increase in northern genera. In fact, it is not a function of unchanged northern genera but of northern genera going south and then evolving rapidly into new forms. Here, we have to go outside equilibrium theory to get an answer.

> During the late Cenozoic, a phase of orogeny [mountain building] beginning about 12 million years ago resulted in a significant elevation of the Andes Mountain range. A major phase of these orogenic movements occurred between 4.5 and 2.5 million years ago with a rise of from 2000 to 4000 meters. The newly elevated Andes served as a barrier to moisture-laden Pacific winds, and a rain shadow was created on the eastern (leeward) side. The southern South American habitats changed from primarily savanna-woodland to drier forests and pampas, and precocious pampas environments and desert and semidesert systems came into prominence at about that time. Many subtropical savanna-woodland animals retreated northward, and new opportunities favoring higher generic diversity arose for those animals able to adapt to these new ecologies.
>
> (p. 1356)

Obviously this is not yet a full explanation. One needs now to show that the northern animals did indeed grab the open real-estate, and why they and not the southern forms were able to do so. But we are well on the way to an explanation – one that blends ecology, evolution, and geology, using forms of understanding that seem no different in principle from those found elsewhere in science.

Systematics

Start with the species question. Organisms seem to fall into fairly well-defined classes, the bottom taxon level of the Linnaean hierarchy (Figure 6.6), like humans (*Homo sapiens*), and house sparrows (*Passer domesticus*), and cabbages (*Brassica oleracea*). What is the

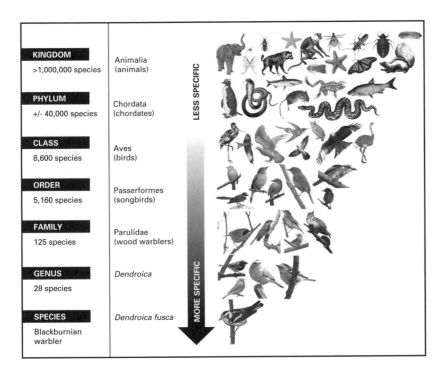

KINGDOM >1,000,000 species	Animalia (animals)
PHYLUM +/- 40,000 species	Chordata (chordates)
CLASS 8,600 species	Aves (birds)
ORDER 5,160 species	Passerformes (songbirds)
FAMILY 125 species	Parulidae (wood warblers)
GENUS 28 species	Dendroica
SPECIES Blackburnian warbler	Dendroica fusca

LESS SPECIFIC

MORE SPECIFIC

Figure 6.6   Linnaean hierarchy

status of these classes? If we have learnt anything, it was that Darwin was ambiguous on this matter. On the one hand, he was proposing a theory of evolution and hence had to argue that life is fluid and that species cannot be anything absolutely permanent or real: species are simply varieties further on down the line. On the other hand, he wrote a book called the *Origin of Species*! He was not writing about something that did not exist. Species are groups that are reproductively isolated. More than this, like everyone else, Darwin recognized that in some sense species seem real or objective, in a way that taxa at other levels (categories) often do not. This paradox has continued to occupy the thinking of evolutionists down to the present (Ereshefsky 1992). Evolution means that everything is in flux. Species are real or objective. How does one solve it?

Well, in cases like these, it is always good to know what you are talking about, so ask first: What is a species? Most obviously, it is a group of organisms that look alike. You and I belong to the same species, *Homo sapiens*, because we both have heads and trunks and arms and . . . etc. etc. We do not belong to the same species as my ferrets (*Mustela putorius*), because they are small and hairy and have paws with claws and . . . etc. etc. You can throw in behavior too, if you like. You and I both read. My ferrets spend much of their time grabbing things and hiding them under the sofa. I do not think that look-alikeness is a silly way of characterizing a species, but there are obviously some problems. If you are a woman or a man then we can start right there. Why not put my wife, Lizzie, and my daughters, Rebecca and Emily, into one group and me and my sons, Nigel, Oliver, and Edward, into another group? Why not put Lizzie and Rebecca and Emily in with the female chimpanzees, and me and the boys in with the males?

Without throwing out the idea of similarity – after all, it is something that we look like one another and not like ferrets – is there another approach we might try? What about interbreeding? Darwin was certainly not the first to pick up on this idea. Species are groups that have members that breed together, and with no one else. Lizzie and I have done our share of breeding. Fond as I am of my ferrets, the union is unconsummated. Of course, even here you have to expand things a bit – I am not in the breeding business with other males – but you get the idea. In fact, among the founders of the synthetic theory, this was a very popular characterization of species. So much so, in fact, that people like Mayr (1969) called it the "biological species concept," implying that it was somehow more special, deeper, than other species concepts, especially any based simply on looks (which was generally known as the "morphological species concept"). Not everyone was a hundred percent happy with the notion, however – especially paleontologists like G. G. Simpson (1962). He introduced the "historical or paleontological species concept," which essentially was the idea of a line of interbreeders through time – like the biological species concept stretched out. Of course, in theory for the evolutionist, the stretching never ends, but in real life the fossil record is sufficiently incomplete that you can turn the gaps to your own ends and pretend that they represent real biological breaks rather than simply the vagaries of successful fossilization. *Homo sapiens* is one species, *Homo erectus* is another.

But why would one want to argue that species are objective or real groupings? People like Mayr sometimes used to argue as though the biological species concept made for objective groups in a kind of self-validating way. They don't interbreed? Then of course they are two groups, two real groups! But this is not entirely satisfactory. Apart from anything else, there is an element of deception involved. Mayr spent the first ten years of his life in America in the bird collections of the American Museum of Natural History. He worked exclusively with skins and, unless he looked out of the window, never even saw a bird courting, let alone breeding. Surely, in some sense, the museum-bound Mayr was doing biology and acknowledging the objectivity of species. Without such assumptions, he could not have written his brilliant and influential *Systematics and the Origin of Species*. But can it really be that morphology is as unimportant as Mayr and others were claiming it to be?

Cutting to the quick, two suggestions have been forwarded to explain the reality of species. One suggestion is that we have been making a major mistake about the nature of species (Ghiselin 1974a; Hull 1980). The Linnaean hierarchy implies that we are dealing with classes or sets – groups whose members have common features. The set of triangles is the group whose members are plane figures with three sides. We have been thinking of species as groups of organisms with common features – characters or breeding abilities or whatever. This is wrong. Truly, species are like organisms – they are integrated entities. In other words, they are individuals. There are connections between humans – relationships – just as one finds between the parts of organisms or between other individuals like countries. It is not just a matter of physical nearness. Alaska is part of the US in a way that Ontario is not, even though the first does not touch another US state and the second does. Michael Ruse and Lizzie Ruse are not members of the class *Homo sapiens*. We are parts of the individual *Homo sapiens*. And just as we would say that Michael and Lizzie Ruse, and the USA, are real things – objective entities – so also we can say that *Homo sapiens* is a real thing – an objective entity.

There are some interesting – critics would say odd – consequences of this way of thinking. Suppose you met someone who was short, fat, stuck his hand in his jacket, spoke French with a Corsican accent, and planned on invading Russia. Nothing would make you think he was really Napoleon. The real Napoleon lies buried in Les Invalides in Paris. Individuals do not repeat. So extinct species cannot repeat. If someone worked out the DNA sequence of *Australopithecus afarensis* or *Tyrannosaurus rex* from various bits of incompletely fossilized bone, and then re-created a being, it would not be a genuine australopithecine or tyrannosaurus. It would have to be a new species. Of course, whether or not this line of argument is right is all a matter of intuitions, but some of us have intuitions that go the other way. Perhaps more troublesome is the very idea that species are all that integrated. When group selection ruled the world, this was a reasonable suggestion. But in the age of individual selection, such talk seems hollow. My wife and I do cooperate, I am glad to say. But I snatched her out from under the noses of several competitors and thought not a bit about their desires – rather enjoyed their envy, actually. Not much integration there, I am afraid.

The other suggestion is more philosophical in a way. William Whewell (1840), of all people, suggested that the reason why we think that certain minerals are real and not just compounds is because in some sense they are the focus of certain law clusters. And more particularly, they show that grouping one way is not arbitrary, because there are independent, other ways of grouping them. In other words, you have connections which show that everything that is one thing is another thing, and everything that is that other thing is the first thing. This is not chance, but a reflection of the way that nature is divided at the joints, to use the phrase of Plato. Actually it is not really such a surprise that Whewell would think this way, because (remember) he started out as a Cambridge professor in a field where classification is important, namely mineralogy. Only later did he move to philosophy. As Herschel pointed out, this approach to classification is linked to Whewell's thinking abut the significance of consilience. There too, different areas of human understanding are connected through the ultimate reality of things. Newton's gravitational force is real because it manifests itself in different dimensions. So a mineral is real because it has both a certain molecular structure and certain physical properties. Something that had no correlation elsewhere would be arbitrary, inessential.

Can you apply this thinking to species? Well, yes you can! By and large, groups delimited morphologically are those very groups delimited by breeding criteria (Ruse 1987). That, as I have said, was exactly the basis of Mayr's work in the 1930s. It was the basis on which the birds that Darwin brought back from the Galapagos were labeled species. The ornithologist to whom Darwin showed them, John

Gould, could tell from the physical features that these were separate, interbreeding groups. Humans look like each other. Humans copulate with each other. Humans do not have sex with ferrets. Humans are a real grouping. This is why species are real. As with the species-as-individuals thesis, there are problems. Humans do not look exactly like each other. So a certain amount of flexibility is needed here. Nevertheless, teasing though it may be to include my wife and daughters in with the female chimps, let me assure you that physically and behaviorally they are a lot more like male humans than they are like any kind of ape. Some groups, as in the ring-of-races situation, cause problems; but probably in such cases there is debate about whether we have real separation at such a point. Are we dealing with one species breaking into two or two species still joined? Also many organisms do not have sexuality. Indeed, if you think at the micro-level, there are so many organisms without sexuality that some biologists have argued (with good reason) that the species problem is really a rather insignificant topic, magnified by the special interest of large-animal biologists like Mayr.

But even if you stay at the macro-level, the simple fact of the matter is that species in biology are a problem for the classifier – not all of the time, obviously, but as soon as you start to think about them theoretically. Which of course is precisely what you would expect. Our thinking apparatus is tuned to deal with groups that separate off. Humans are potential threats or mates. Tigers are threats. Artichokes are good all the way down. Evolution over time and space is not something our ancestors encountered. We tend to make mathematical objects like triangles – things with clean and clear definitions – as our paradigms of objects of classification. Humans, tigers, artichokes are treated in the same way. So, if Darwin is right and you have to think of groups as fluid, non-permanent, there is bound to be a breakdown. The species problem is a fascinating problem, but in the end it may be a waste of time to try to force species to be more than they are. They are objective, but only so far. Leave it at that.

Cladism

What about classification generally? Because it explained why the Linnaean hierarchy seems to work, Darwin was pleased with his system.

The way that organisms fall into nested sets reflects the fact that they have evolved, and the closer organisms are in the hierarchy the closer these organisms are in evolutionary history. But can one show all of this in a more systematized sort of way? In the past three or four decades, there has been something of a revolution in systematics by people who think that this is possible (Ridley 1986; Hull 1988). These people – known as "phylogenetic systematists" or "cladists" – are followers of a mid-twentieth-century German taxonomist, Willi Hennig. They argue that the aim of taxonomy is to reflect phylogeny only: so, even if (say) crocodiles and birds are totally different, if they are more closely related to each other in the sense that their shared ancestor occurred more recently than did the last shared ancestor of any other species, the classification must reflect this. If two groups evolved in parallel or converged on similar solutions (both mammals and birds are hot blooded), this is interesting but irrelevant to classification.

How is classification to be done, given that the fossil record is usually missing and always incomplete? One does it through comparisons of living organisms, combined with hypotheses about how they might have evolved. Suppose you have three different organisms, A, B, and C. A and B share a feature, let us say feathers, whereas C does not – it has scales just like the joint ancestor of all three organisms, D (Figure 6.7). What are the possible phylogenies? You could have had A and B gaining the feathers early and then splitting apart. Or you could have had A going off alone early and getting feathers, and B and C staying together and then splitting when B got feathers (independently of A) and C did not. At this point you bring in Occam's razor. What is the simplest hypothesis? You appeal to the principle of parsimony. Clearly it is simpler to have had just one move to feathers rather than two, in other words the first scenario is preferred over the second.

You keep doing this with different features and seeing what kinds of diagrams of connection (known as "cladograms") you come up with. Then you put your results together, seeing if they fit consistently, to get your final result. One important trick is that of outgroup comparison. Suppose you have three organisms, E, F, and G, where F and G seem more similar to each other than to E (say a human and two dogs). Why put F and G together – a move that is intended to reflect ancestry rather than mere similarity – rather than E and F together? You take another organism H, like the first three but clearly more different than any – say a snake. The common ancestor of E, F, G, and

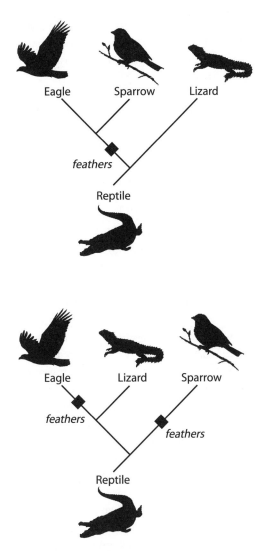

Figure 6.7 Cladograms used to establish classifications

H is farther back in time than any other common ancestors. Now to get to E, F, and G, you need change from the common ancestor to E, F, and G; but if you put F and G together then you need one change only to get F and G different from E (in our case to get one set of human-like features and one set of dog-like), whereas if you put E and F together and G separate, you need two moves to get doggy features. It is more parsimonious to group F and G – but note, again, this is intended to reflect ancestry rather than just similarity.

Or is it? Cladism raises a plethora of philosophical problems, starting with the suspicion that this might all be smoke and mirrors. You claim to be talking about ancestry, but really you are just talking about similarities and differences. Is cladism really an evolutionary system of classification at all? Is it not in fact pre-evolutionary – something that could have been done, that was done, by men like Louis Agassiz, and that really ignores the Darwinian revolution completely? My sense is that this position – sometimes called "transformed cladism" – is not really so silly. Indeed, I think that there is a lot of truth to it. But I do not find it at all worrying. The point is not whether the practice of cladism requires evolutionary principles. The point is whether we can give it an evolutionary interpretation. Which we obviously can. Moreover, if not evolution then what? The possibility of classification becomes inexplicable at best and a miracle at worst, in the sense that miracles are the very last thing that even the most devout scientist wants in his or her science.

Perhaps more worrying is the principle of parsimony (Sober 1988). Why should the world be simple? The answer is that there is no reason. It is simply that it is easiest for us to think that way. KISS – keep it simple, stupid. Cladists tend to be great enthusiasts for the philosophy of Karl Popper, arguing that simplicity means more testable and hence more open to falsification. I am not quite sure that any of this is so, or even that falsifiability is always the best mark of science, but I would agree that simplicity is certainly important in science, for whatever reason. However, I qualify this agreement with the cavil that I am not entirely convinced that simplicity is the only thing at stake here. Let us move on from the question of an evolutionary interpretation of cladism to a Darwinian interpretation. The relationship between cladism and Darwinism is tense, which is hardly surprising, given that cladism was developed by an evolutionist from a tradition that tended not to be very Darwinian. In many respects, cladism is, if not anti-Darwinian, then not very pro-Darwinian. For a start, the only changes that can be recorded by the method are those that occur when species divide into two: there is no place for changes from one classifiable group to another that occur in a single line without splitting. It is not that these do not occur or that they are significant or insignificant. It is just that the method ignores them. Such nonsplitting changes may never occur, but there is nothing in Darwinism that tells us this and much that suggests that it might not always be true. Likewise, a three-way split must be analyzed in terms of two two-way splits. And of course, the method ignores adaptation. Two organisms being brought together by similar adaptive pressures is irrelevant. As is traditionally the case, taxonomists find adaptation a nuisance, because it conceals what they regard as deeper connections.

However, it does seem that the cladists' notion of simplicity is not entirely disconnected from Darwinism. Basically, Darwinism suggests that change of independent lines in the same direction is not at all impossible, but not necessarily very likely. In other words, if you have two independent organisms sharing the same features, it is more likely to be because they inherited them than because they developed them independently in the same way. It is not just that the first explanation is simpler. It is what Darwinian theory, combined with experience, suggests is a better biological bet. If one had a different causal theory – for instance, one that suggested that change in newly separated organisms will generally go the same way for several generations – then it would not be a question of simplicity to think that one should prefer the hypothesis of inheritance from a shared feature, but of being wrong. Perhaps part of the trouble here is that, as already mentioned, cladists tend to be followers of Popper and he did not much care for Darwinism and cared even less for making hypotheses on the evidence of past examples. That, for him, was a matter of inductive generalization – going from several examples to new unseen cases – and because it was not deductive, it was therefore WRONG. But we need not be hung up in quite this way. I suspect that there is more Darwinism behind cladism than most of its practitioners care to admit.

Cladists tend to take credit for all that has been done on classification and the working out of phylogenies in the past few decades. That is not quite fair. Molecular biology has played a major role, allowing comparisons at the level of protein and nucleic acid. Particularly important has been the notion of a molecular clock, which tries to assess rates of evolution through the changes that are believed to be the result of non-adaptive drift (Hillis, Huelsenbeck, and Cunningham 1994). Also important in working out phylogenies has been our increased knowledge of the fossil record and the ability to assign absolute dates. And let us not forget the power of computers in grinding up massive amounts of data. But whoever should get the credit, it is

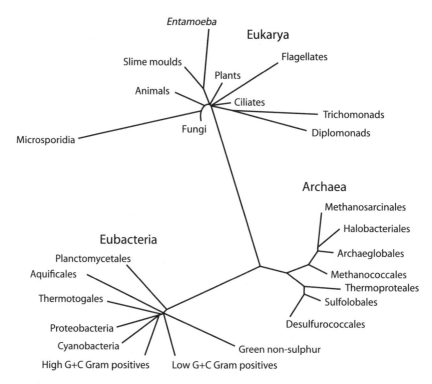

Figure 6.8 The history of life

certainly true that now we have far more understanding of life's history and diversification than they had fifty years ago, let alone at the time of Darwin. Figure 6.8, based on some of the best data of recent time, shows, for instance, that the traditional division of organisms into animals and plants is a very small part of the overall picture, and that much more significant overall for the history of life were some of the earlier divisions, particularly that of living organisms into prokaryotes and eukaryotes. Darwin had little to say about the course or path of evolution. Now we know very much more.

Morphology

For Darwin and his contemporaries, the big issue was form and function, or rather: form or function? Which is the more important? Is function something that creates and drives form, or is form more basic, in which case function simply fits in with what is allowed? It is this issue that we live with today. Since generally function is spelt out in terms of natural selection (although more on this assumption in a moment) the big question is whether the world of organisms is adapted through and through, by the activity of selection. I stress one thing as we start this discussion. No working evolutionist today is going to deny that the world is greatly marked by adaptation, by functioning characteristics, and these are the result of natural selection. Eyes, teeth, noses, flippers, paws, wings, leaves, flowers, roots – these are adaptations and selection was the key causal factor. To a great extent, this is the working hypothesis that forms the background for the evolutionist. It is not an *a priori* assumption – we have seen in earlier sections strong reasons to think that selection does bring on adaptation – but is more like the null hypothesis, against which one tests the work that one is doing, and which provides the laser beam with which one explores the world of life.

This is not to say that evolutionists are stuck with the assumption that things are adaptive and simply leave matters at that. One wants to find out exactly how things function and why. A number of tricks or methods are important here (Ruse 2003). First, one might run a comparative study. By comparing organisms across a range, one can work out what is happening, or test hypotheses. Much of this is so obvious one might not even realize what is going on. Take the extinct, saber-toothed tiger, a brute with massive downward-pointing fangs. What is the function of the fangs? Attacking its prey and enemies, obviously. We know that that is what animals use their fangs for today and so we argue analogously. In fact, we can probably go a bit further and, by looking at the fangs and making comparisons, say whether they are for stabbing or shearing or whatever. Apparently, given the size and shape of the fangs, and their serrated edges, stabbing is the most likely hypothesis. Since the tigers were not built for fast pursuit, it is thought that they ambushed their prey and went for the soft spots, bleeding their victims to death.

Taking a broader perspective, one might compare several species from the same group. One well-known study is of testis size (Harcourt, Harvey, Larson, and Short 1981). Why do some animals like the chimpanzees have huge testicles given their size, and other animals – sometimes closely related, like the gorillas – have only small testicles? Because of the differences and the relationships, it is unlikely that it is simply a matter of shared inheritance from earlier forms. Selection has been at work. Sperm competition is important here. If a female is mating with several animals when she is in heat, then (from a reproductive perspective) a male having lots of sperm is going to be at an advantage – better a teacup than an eggcup. The sheer stupidity of individual selection dictates that, although competition could be just as intense at the eggcup level, and although ultimately there must be a balance between cost and payoff, if one gormless copulator is going the teacup route, then everyone must. And as it happens, this seems to be the case, at least among the primates. Gorillas have harems and only the dominant male gets to breed with the fertile females. He has no need to produce lots of sperm. Chimpanzees live in groups and females are receptive to many males. Sperm production is at a premium. Apparently humans come somewhere in between, although compared to other animals we are very far behind. Think of that little bird, the dunnock or hedge sparrow, who has a complicated sex life like something out of the *Karma Sutra*, with (among other combinations) several females mating with one male. Were he scaled up to human size, he would have testicles the size of mangoes.

Another way of ferreting out adaptation is by "reverse engineering". You look at the feature, and then work backwards to try to understand its design or function. A classic case is that of the strange diamond-shaped plates that run along the back of the dinosaur Stegosaurus (Figure 6.9). What were these for? Several hypotheses have been proposed. The most obvious is that the plates were used for fighting, either attack or defence. However, detailed study of the bones shows that they simply were not strong enough for combat: they would break and crush far too easily. Another hypothesis is that the plates were used for sexual attraction, but again there are problems. There does not seem to be the dimorphism (difference between the sexes) that one associates with sexual selection. The popular hypothesis today is that the plates were used by the cold-blooded animals to regulate their heat, catching the sun in the morning and then letting heat be blown away later in the day when the sun was hotter or heat was produced by digestion (Farlow, Thompson, and Rosner 1976). The plates look just like the fins used in power plants to transfer heat. Also, one can show that the blood seems to have been carried through the plates in just such a fashion as would maximize heat transference. And

Figure 6.9 Stegosaurus

finally, tests on models confirm that the Stegosaurus was built in just such a fashion as to support the temperature-control hypothesis. All in all, a neat piece of backwards reasoning and then testing to show the plausibility of the hypothesis.

A third way of trying to work out adaptation is through so-called "optimality models". Here, one tries to offer a theoretical argument about how resources would be best used, and then show that this is what is happening in the real world (Orzack and Sober 2001). "In order to employ engineering optimization models the biologist tries to interpret living forms as in some sense the 'best.'" What does one mean by "best"? "In effect the biologist 'plays God': he redesigns the biological system, including as many of the relevant quantities as possible and then checks to see if his own optimal design is close to that observed in nature" (Oster and Wilson 1978, 294–5). Some philosophers have been very negative about these modes of approach, arguing that (in the words of Bertrand Russell about something else) they are simply "postulating what we want," and that, while this practice "has many advantages," regretfully "they are the same as the advantages of theft over honest toil" (Brandon and Rauscher 1996, 200). But this is surely a grotesque exaggeration. If evolutionists did nothing but postulate, there would be legitimate cause for complaint. But the method described demands significant empirical investigation, often working across groups, to see why there are different forms and in what ratios. Edward O. Wilson himself has done this very successfully when studying the forms of the ants and their ratios – why so many soldiers, why so many gatherers, why so many nurses, and so forth. The ant genus *Atta* has seven classes of workers.

> A key feature of *Atta* social life . . . is the close association of both polymorphism and polyethism with the utilization of fresh vegetation in fungus gardening . . . An additional but closely related major feature is the "assembly-line" processing of the vegetation, in which the medias cut the vegetation and then one group of ever smaller workers after another takes the material through a complete processing until, in the form of 2-mm-wide fragments of thoroughly chewed particles, it is inserted into the garden and sown with hyphae.
>
> (Wilson 1980, 150)

Why the different castes and why the proportions? "What *A. sexdens* has done is to commit the size classes that are energetically the most efficient, by both the criterion of the cost of construction of new workers . . . and the criterion of the cost of maintenance of workers." Similar reasoning occurs when optimality theorizing is used to explain the nature of a particular adaptation. Why for instance do praying mantises always attack prey of a particular size? Simply because this is the most energy efficient, in the sense of the best ratio of energy used to energy received. If the prey were bigger then the mantis's limbs could not do such a good job and if the prey were smaller then the mantis would be using more energy than needed for the prey that is captured.

Spandrels

Is the Darwinian committed to absolutely everything being constantly adapted? Stephen Jay Gould was one who argued that Darwinism overdoes the reliance on adaptation. He claimed that much in the organic world is really spandrel-like. These are the triangular planes between arches in medieval buildings, often – as in the case of the church of San Marco in Venice – with beautiful decorations all over them (Figure 6.10). "The design is so elaborate, harmonious, and purposeful

Figure 6.10 Spandrel from the Villa Farnesina, Rome, painted by Raphael and co-workers

that we are tempted to view it as the starting point of any analysis, as the cause in some sense of the surrounding architecture." But this is to reverse cause and effect. "The system begins with an architectural constraint: the necessary four spandrels and their tapering triangular form. They provide a space in which the mosaicist worked; they set the quadripartite symmetry of the dome above" (Gould and Lewontin 1979, 148). Perhaps, argued Gould, much of the living world is non-adaptive,

Figure 6.11 How the elephant got its trunk (Rudyard Kipling, "Elephant's Child", *Just So Stories*, 1902)

for one reason or another, and we should recognize this. He sneered that those who overdo adaptation are like Dr. Pangloss in Voltaire's *Candide*, forever seeing value or use when there is none – "the best of all things in the best of all possible worlds." He ridiculed Darwinians, accusing them of spinning "just so" stories, akin to the fantastic tales of Rudyard Kipling who supposed (among other things) that the elephant has a long trunk because a crocodile pulled it (Figure 6.11)!

Gould was a master of rhetoric. Although no one would deny that sometimes evolutionists do get carried away with enthusiasm for their theories, the force of his charge is another matter. The background, as has been stressed already, is that there is massive evidence for the ubiquity of adaptation and that Darwinians are not being foolish or non-scientific in expecting complex or strange features to have been produced by selection. Things like the Stegosaurus's plates do not just happen by chance. The Catholic Church accepts miracles but it is not being foolish or non-religious in expecting to find natural causes for strange and wonderful cures. This is the null hypothesis, the background assumption. The Church does, then, accept miracles if it feels forced to, and likewise Darwinians have always accepted that not all features are going to be adapted. They assume adaptation but then qualify and except if they feel forced to. Already we have seen that drift can cause the non-adaptive and that homology is non-adaptive. Also, in cases like balanced, superior-heterozygote fitness, nature plays off the fit against the unfit. A person with two sickle-cell genes is certainly not adapted. Realize also that what might have been adaptive once is not necessarily adaptive now. The human appendix, for example, and the four-limbedness of vertebrates. The earliest vertebrates were aquatic, and two limbs fore and two limbs aft were needed to rise up and down in the water – just like planes today which use two wings forward and then tiny winglets at the back to rise up and down in the air (Maynard Smith 1981).

Sometimes things have changed so rapidly that adaptation has simply not yet caught up. The dunnocks are parasitized by cuckoos, which lay eggs in their nests. Other species of bird in England that are parasitized in this way have great abilities to detect strange eggs, and the cuckoos have responded by producing eggs that look like those of the hosts. Not so for the dunnocks. Why? Almost certainly because the dunnocks are only recently parasitized and hence selection has not brought on detection abilities in them or countermeasures by the cuckoos. Then there are cases where natural and sexual selection conflict. The peacock's tail may excite the peahens, but it is nothing but an encumbrance when the peacock is fleeing from predators. More complexly, we have the case of the Irish elk with its massive antlers (Figure 6.12). Were these adaptive? Possibly not, given their size and the food needed to produce them each year. But if the antlers grew bigger each year – perhaps growing even faster than body size (this is known as "allometric" growth) – then young males were able to attract females, even though as full adults they would be at a disadvantage. By this time, however, the damage was done because their genes had been passed on. Finally, let us note that sometimes genes have multiple effects (they are "pleiotropic"). It might be better to have the good effects even though there are also bad effects. And, analogously, sometimes the engineering does not work that well.

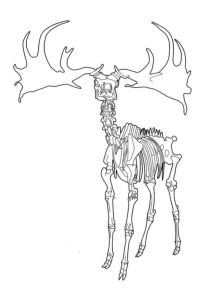

Figure 6.12 Skeleton of the Irish elk

Humans have big heads and walk upright – both good things. However, women suffer and sometimes die in childbirth, because big heads require large birth canals and walking upright puts serious constraints on how wide the hips can be.

Organized complexity

No Darwinian is saying that everything is necessarily adaptive all of the time. But what about natural selection? Is it necessary to assume always that it is the cause of adaptation? Since selection is not a tautology, it is not the case that it is logically and necessarily the cause of adaptation, although some thinkers – British biologist and popular science writer Richard Dawkins (1983) particularly – have argued strenuously that it is, as a matter of empirical fact, always the cause. He claims that adaptive complexity demands a solution, and that blind law does not and cannot produce it. In the real world, things go wrong not right. Planes crash and break up. They do not spontaneously reassemble. Lamarckism – the inheritance of acquired characteristics – would be one plausible cause of adaptation, but modern genetics shows that it is wrong. Saltationism, evolution by jumps, simply cannot lead to functioning complexity. And there are no other options in the natural world. It is selection all of the way.

Recently, a number of physicists and fellow travelers have mounted a counter-argument. They argue that Darwinians are too quick in supposing that blind law cannot lead to adaptive complexity. Perhaps the laws of physics and chemistry unaided can do the job. One who used to argue this way at the beginning of the last century was the Scottish morphologist (and great favorite of Gould) D'Arcy Wentworth Thompson (1948). He thought that much in the living world was simply a matter of physical law.

> Cell and tissue, shell and bone, leaf and flower, are so many portions of matter, and it is in obedience to the laws of physics that their particles have been moved, molded and conformed. . . . Their problems of form are in the first instance mathematical problems, their problems of growth are essentially physical problems, and the morphologist is, *ipso facto*, a student of physical science.
>
> (p. 10)

Hence: "We want to see how, in some cases at least, the forms of living things, and of the parts of living things, can be explained by physical considerations, and to realize that in general no organic forms exist save such as are in conformity with physical and mathematical laws" (p. 15). A cherished example was that of the jellyfish, which has a form identical to that of a drop of ink falling through water.

> The living medusa has a geometrical symmetry as marked and regular as to suggest a physical or mechanical element in the little creature's growth and construction. . . . It is hard indeed to say how much or little all these analogies imply. But they indicate, at the very least, how certain simple organic forms might be naturally assumed by one fluid mass within another, when gravity, surface tension and fluid friction play their part, under balanced conditions of temperature, density and chemical composition.
>
> (pp. 396–8)

Sometimes, to be candid, it was not easy telling if Thompson thought that forms had adaptive values that physics produced, or if he (and Gould and others following him) simply thought that the whole question of adaptation was overblown and irrelevant. Many of his successors, however, are believers in adaptation, thinking that selection is not needed. Self-organization is the key. Physics and chemistry, in the felicitous words of the American computer enthusiast Stuart Kauffman (1993), yield "order for free."

A perennial favourite is the pattern one finds in many flowers and seeds – a kind of diamond crisscross, known as "phyllotaxis." It is shown clearly in the way that the seeds in a sunflower are packed, and no less clearly in the patterns shown on the outside of pinecones. Why is there this pattern, which analysis shows is governed by a certain interesting arithmetical series – the Fibonacci series (of *Da Vinci Code* fame)? Anti-Darwinians argue that these patterns are a function of the ways in which things grow according to the laws of mathematics and that they have no connection with selection or adaptation. They just are (Goodwin 2001). In more detail, phyllotaxis, is produced by the leaves or analogous plant parts appearing at the centre (the "growing apex") and then being pushed outwards (Mitchison 1977). The leaves or parts follow a path known as the "genetic spiral." This forms the crisscrossing diagonal spirals, known as "parastichies." A Fibonacci series is formed by adding together the previous two members of the series, starting with zero and one. Thus one has 0, 1, 1, 2, 3, 5, 8, 13, . . . , or, more generally, $n_j = n_{j-1} + n_{j-2}$. For any particular plant species, the numbers of the parastichies, one set clockwise and one set counterclockwise, are always related by being consecutive numbers of the Fibonacci series. In Figure 6.13, the example is of an 8, 13 phyllotaxis.

But what does this all signify? For Darwinians, the answer is obvious. We have something brought on by natural selection for adaptive ends. Thus, Chauncey Wright, a philosophical pragmatist and keen Darwinian in Boston a decade after the *Origin*, wrote: "To realize simply and purely the property of the most thorough distribution, the most complete exposure to flight and air around the stem, and the most ample elbow-room, or space for expansion in the bud, is to realize a property that exists separately only in abstraction, like a line without breadth" (Wright quoted in Gray 1881, 125). For the self-organization supporters, the answer is not so simple. "The tapestry of life is richer than we have imagined. It is a tapestry with threads of accidental gold, mined quixotically by the random whimsy of quantum events acting on bits of nucleotides and crafted by selection sifting. But the tapestry has an overall design, architecture, a woven

Figure 6.13 Pine cone: an example of an 8, 13 phyllotaxis

cadence and rhythm that reflect underlying law – principles of self organization" (Kauffman 1995, 185). Canadian-born morphologist Brian Goodwin argues that the patterns we see in phyllotaxis are completely and utterly explained by mathematics – they follow as a function of lattice theory. There is no need to invoke selection at all. They work just fine, but not because of Darwinian factors.

This is an ongoing debate. As you might imagine, Darwinians are not silenced by the self-organization crowd.

> Computer simulations indicate that phyllotaxy can influence the quantity of light intercepted by leaf surfaces. Model plants constructed with equal total leaf area and number differ significantly in flux, even when leaf-divergence angles are very similar. . . . Nonetheless, computer simulations indicate that a variety of morphological features can be varied, either individually or in concert, to compensate for the negative aspects of leaf crowding resulting from "inefficient" phyllotactic patterns. Internodal distance and the deflection ("tilt") angle of leaves can be adjusted in simulations with different phyllotactic patterns to achieve equivalent light-interception capacities.
>
> (Niklas 1988, 566)

Fortunately, we can leave the debate to the contestants. Our survey of the field is finished. In recent years, morphology has been one of the most interesting and contentious members of the evolutionary family. One suspects that if Darwin were still alive, he would be in the thick of the discussion.

Embryology

Finally, we come to embryology, or what today is known as evolutionary development ("evo-devo"). Fifty years ago this would have been a very short section – a paradox, for, as we know, Darwin thought embryology very important, even claiming it was the most important evidence that he had. But, as we know also, after the *Origin* embryology became a crucial part of the second-rate, Germanic, phylogeny tracing that swamped evolutionary thinking. Embryology was certainly not welcoming of selection studies. Nor did things change much at the beginning of the twentieth century, when genetics was starting to establish itself. Now embryology had turned more experimental and its enthusiasts were claiming that they alone had the keys to life's mysteries. It is hardly surprising, therefore, that when the synthetic theory was put together – a move combining selection and genetics – those at the cutting edge tended to ignore embryology. Why should they embrace something that was still so dismissive of the work they thought crucial? For better or for worse – and one can argue that there were good reasons for thus simplifying the crucial evolutionary models – synthetic theorists treated the transition from gene to finished organism as essentially unimportant.

Things have changed. In the past two or three decades, embryology has really picked up again, as molecular biologists have become interested in development, tracing the journeys from the DNA to the finished organism (Carroll 2005; Carroll, Grenier, and Weatherbee 2001). And what incredible, unexpected findings they have unearthed! We have seen the homologies between the forelimbs of vertebrates. Who would have dreamed that there would be molecular homologies linking organisms as different as humans and fruitflies? Forty years ago, Mayr denied explicitly that organisms this far apart could be homologous in any way. But he was wrong. Truly incredible, if they did not actually exist, would be certain so-called "homeotic genes." These are the genes that regulate the identity and order of the parts of the body. It is they that tell the body where to put an eye and where to

The Consilience: Two

put a leg. Something going wrong with one of them might perhaps move an eye to where a leg might normally appear, or vice versa. Among such homeotic genes there is a subclass – found in bilaterans (organisms that are the same on both sides) – the so-called "*Hox* genes." It is these that order the appearance of various bodily parts, and perhaps expectedly they work in the sequence that they are found on the chromosomes. In fruitflies, *Drosophila*, the *Hox* genes begin with the head, continue down through the thorax, and finish at the end of the abdomen. The *Hox* genes have comparable functions in vertebrates including humans. Within these genes, one finds identifiable lengths of DNA that play specific, vital functional roles in this process. What was truly astounding was the discovery of homology between the DNA lengths of *Drosophila* and other bilaterans, from frogs through fish and mice to humans (Figure 6.14). The flies' legs and the humans' legs go back to the same processes.

We have here the strongest possible evidence for evolution. Anyone who says that our thinking about evolution today is not scientific is simply speaking from ignorance or prejudice. We have just the sort of phenomenon that Whewell identified as the defining mark of the best kind of consilience. Does this speak also to the theory of natural selection? Some enthusiastic workers think not.

> The homologies of process within morphogenetic fields provide some of the best evidence for evolution – just as skeletal and organ homologies did earlier. Thus, the evidence for evolution is better than ever. The role of natural selection in evolution, however, is seen to play less an important role. It is merely a filter for unsuccessful morphologies generated by development. Population genetics is destined to change if it is not to become as irrelevant to evolution as Newtonian mechanics is to contemporary physics.
>
> (Gilbert, Opitz, and Raff 1996, 368)

Can this possibly be true? Certainly, the whole new science of evo-devo shows that we have to be careful in the assumptions we might be inclined to make about development. It is clear that evolution is a tremendous re-user of what is already available, putting the same things to work in all sorts of different organisms. If it functions for one, then let's use it for others. We find also at the molecular level that we get many instances of what in the nineteenth century was known as "serial homology," where the same part (for example, a vertebra) is used again and again in the same organism. Time and again at the

(a)

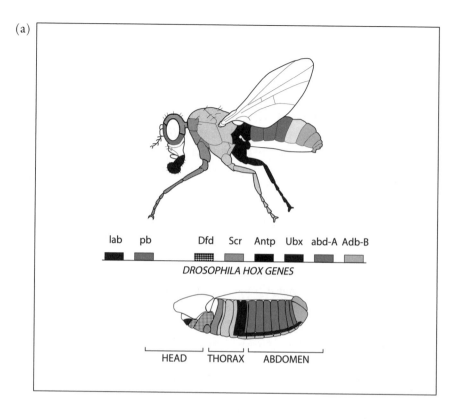

lab   pb   Dfd   Scr   Antp   Ubx   abd-A Adb-B

DROSOPHILA HOX GENES

HEAD   THORAX   ABDOMEN

(b)

Fly Dfd	PKRQRTAYTRHQILELEKEFHYNRYLTRRRRIEIAHTLVLSERQUKIWFQNRRMKWKKDN	KLPNTKNVR
AmphiHox4	TKRSRTAYTRQQVLELEKEFHFNRYLTRRRRIEIAHSLGLTERQIKIWFQNRRMKWKKDN	RLPNTKTRS
Mouse HoxB4	PKRSRTAYTRQQVLELEKEFHYNRYLTRRRRVEIAHALCLSERQIKIWFQNRRMKWKKDH	KLPNTKIRS
Human HoxB4	PKRSRTAYTRQQVLELEKEFHYNRYLTRRRRVEIAHALCLSERQIKIWFQNRRMKWKKDH	KLPNTKIRS
Chick HoxB4	PKRSRTAYTRQQVLELEKEFHYNRYLTRRRRVEIAHSLCLSERQIKIWFQNRRMKWKKDH	KLPNTKIRS
Frog HoxB4	AKRSRTAYTRQQVLELEKEFHYNRYLTRRRRVEIAHTLRLSERQIKIWFQNRRMKWKKDH	KLPNTKIKS
Fugu HoxB4	PKRSRTAYTRQQVLELEKEFHYNRYLTRRRRVEIAHTLCLSERQIKIWFQNRRMKWKKDH	KLPNTKVRS
Zebrafish HoxB4	AKRSRTAYTRQQVLELEKEFHYNRYLTRRRRVEIAHTLRLSERQIKIWFQNRRMKWKKDH	KLPNTKIKS

Figure 6.14 Molecular homologies. (a) In the center are the genes that regulate development, in the order in which they occur on the chromosome: the fruitfly develops, under the influence of the genes, from larva (*bottom*) to adult (*top*). (b) The information that regulates the development of the fruitfly is almost identical with the information that regulates the development of the human, the frog, the chick, the mouse, and others. On the left is a list of the genes in the different organisms that regulate the production of a key substance (a protein) needed for development. This substance – protein – is a chain of smaller units – amino acids – represented by capital letters in the rows on the right. It can be seen that this protein, called a "homeobox," is virtually identical in organisms of very different kinds

156   *The Consilience: Two*

molecular level, instead of inventing a new gene or a new process or a new structure, existing genes or processes or structures are duplicated or triplicated or more, and then put to use in the same or slightly different ways.

But surely, rather than calling for the rejection (or redundancy) of selection, this enriches the existing picture. For all that Charles Darwin virtually defined himself against *Naturphilosophie*, he plundered their discoveries to create his own biological world picture. Serial homology may have been a key feature in the idealists' *Weltanschauung*, but Darwin was happy to make use of it, stressing that natural selection is a process that works with what is at hand, modifying and multiplying, rather than creating anew. This was ever a key theme of the great English naturalist. In a little book on orchids, written just after the *Origin*, Darwin stressed that if something is needed, selection cannot simply design from scratch. It must work with what is available, and this will often lead to duplication and like phenomena, until the end product looks more like a Rube Goldberg (in England, Heath Robinson) contraption, than anything designed by an intelligent being.

> Although an organ may not have been originally formed for some special purpose, if it now serves for this end we are justified in saying that it is specially contrived for it. On the same principle, if a man were to make a machine for some special purpose, but were to use old wheels, springs, and pulleys, only slightly altered, the whole machine, with all its parts, might be said to be specially contrived for that purpose. Thus throughout nature almost every part of each living being has probably served, in a slightly modified condition, for diverse purposes, and has acted in the living machinery of many ancient and distinct specific forms.
>
> (Darwin 1862, 348)

In short, the Darwinian welcomes the new discoveries. Never think that challenges are threatening. Darwinian theory is a vigorous, forward-looking part of science. It is not complete, and almost daily there are extensions and revisions. This is a mark of strength not weakness. The best science gives its researchers more problems at the end of the day than there were at the beginning. The best science is dynamic, not static; exciting not boring; challenging not easy. On all of these counts, Darwinism today earns a high grade.

7

Humans

Human beings are part of the living world. They are the product of evolution. Through their culture, humans have become organisms of a very distinctive and (especially to us) interesting kind. But ultimately, for the Darwinian, humans are the result of selection, natural and sexual. There can be no compromise on this. If one believes in God and the possibility of miracles of one sort or another, they cannot enter into the discussion of the origins and nature of humankind. Notions like the soul, even if they make sense and are real things, are not part of the scientific discourse. Nature must stand on its own.

The Descent of Man

On the basis of our anatomy and development, Charles Darwin started *The Descent of Man*, his work on our species, by tying us firmly into the animal world.

> It is notorious that man is constructed on the same general type or model with other mammals. All the bones in his skeleton can be compared with corresponding bones in a monkey, bat, or seal. So it is with his muscles, nerves, blood-vessels and internal viscera. The brain, the most important of all the organs, follows the same law, as shown by Huxley and other anatomists.
>
> (Darwin 1871, 1: 10)

There can be only one explanation of this fact.

The homological construction of the whole frame in the members of the same class is intelligible, if we admit their descent from a common progenitor, together with their subsequent adaptation to diversified conditions. On any other view, the similarity of pattern between the hand of man or monkey, the foot of a horse, the flipper of a seal, the wing of a bat, &c., is utterly inexplicable.

(p. 31)

Moving on to our mental powers, bluntly Darwin stated that "there is no fundamental difference between man and the higher mammals in their mental faculties" (p. 35). Adding:

Of all the faculties of the human mind, it will, I presume, be admitted that Reason stands at the summit. Few persons any longer dispute that animals possess some power of reasoning. Animals may constantly be seen to pause, deliberate, and resolve. It is a significant fact, that the more the habits of any particular animal are studied by a naturalist, the more he attributes to reason and the less to unlearnt instincts.

(p. 46)

This was backed by discussion of such topics as language and tool use, which Darwin felt could be seen as existing in rudimentary forms in the animals and hence did not set us humans apart from our evolutionary heritage.

How has this all come about?

As man at the present day is liable, like every other animal, to multiform individual differences or slight variations, so no doubt were the early progenitors of man; the variations being then as now induced by the same general causes, and governed by the same general and complex laws. As all animals tend to multiply beyond their means of subsistence, so it must have been with the progenitors of man; and this will inevitably have led to a struggle for existence and to natural selection.

(p. 154)

Finally, from where and what did Darwin think we evolved? We are more like old-world monkeys than new-world ones, and given the present-day habitats of our closest relatives, the gorillas and chimpanzees, Africa got the prize.

We are naturally led to enquire where was the birthplace of man at that stage of descent when our progenitors diverged from the Catarhine stock? The fact that they belonged to the stock clearly shews that they inhabited the Old World; but not Australia nor any oceanic island, as we may infer from the laws of geographical distribution. In each great region of the world the living mammals are closely related to the extinct species of the same region. It is therefore probable that Africa was formerly inhabited by extinct apes closely allied to the gorilla and chimpanzee; and as these two species are now man's nearest allies, it is somewhat more probable that our early progenitors lived on the African continent than elsewhere.

(p. 199)

Darwin made it clear that he did not think that we are descended from monkeys or apes living today, although this did not stop every cartoonist of the day picturing us humans as one stop from the orangutans (Figure 7.1).

Sexual selection

The Descent of Man is a different book from the *Origin*. Although Darwin wrote the earlier book in an easy style and although he used facts found by many others, one senses always that this was a man who was his own master, in charge of the information and the argument. In the *Descent*, Darwin was already starting to lag behind the work of others, and much that he offered was a digest of, and commentary on, what others had done. From before the publication of his *Evidence as to Man's Place in Nature* in 1863, Thomas Henry Huxley and others had been working on humans and their evolution – not to mention their cultures – and Darwin picked up and synthesized their work. Also (for all that he had originally intended otherwise) Darwin was much more ready in the *Descent* to fall in with the popular, almost religion-substituting, nature of evolutionary thinking as it had grown and developed after the *Origin*. In the *Origin*, Darwin tried to stay strictly away from value judgments. In the *Descent*, there was nothing like this careful (implicit) acknowledgment that the job of science is to describe and explain, no suggestion that prescription is to be left to others. In the *Descent* there was, for instance, approbation of capitalism, with great beneficiaries like Darwin himself given a pass from the cares and trials of ordinary people.

Figure 7.1 Darwin as an ape (*The Hornet*, March 22, 1871)

Man accumulates property and bequeaths it to his children, so that the children of the rich have an advantage over the poor in the race for success, independently of bodily or mental superiority. On the other hand, the children of parents who are short-lived, and are therefore on an average deficient in health and vigour, come into their property sooner than other children, and will be likely to marry earlier, and leave a larger number of offspring to inherit their inferior constitutions. But the inheritance of property by itself is very far from an evil; for without the accumulation of capital the arts could not progress; and it is chiefly through their power that the civilised races have extended, and are now everywhere extending their range, so as to take the place of the lower races.

(Darwin 1871, 1: 169)

Humans 161

The one thing that did distinguish Darwin from all of the others was his commitment to selection, and (as mentioned earlier) it was in the *Descent* that sexual selection really came into its own. As explained, this was spurred by Wallace's rejection of natural selection as an adequate cause of human evolution. Darwin could, of course, simply have turned his back on Wallace's speculations, arguing that since they were fueled by Wallace's turn to spiritualism, they did not merit a scientific answer. But Darwin thought that Wallace had a good point in highlighting such phenomena as human intelligence and human hairlessness as being difficult to explain by natural selection. So, whatever the motive, Darwin felt he had to give a response, and the response was framed in terms of the importance of sexual selection.

Hence, in typical Darwin fashion, the *Descent* included a large – a massive – overview of sexual selection in action, and then this was tied directly to human evolution. For instance, like a lot of mammals, human males are larger than human females. It is clearly established that this kind of sexual dimorphism is a direct result of males competing for females, and is often associated with harems or polygamy (more accurately polygyny, where one male has more than one female). "The gorilla seems to be a polygamist, and the male differs considerably from the female; so it is with some baboons which live in herds containing twice as many adult females as males" (1: 266). There was no way that a Victorian gentleman was going to suggest outright that polygyny is a natural, universal human state, but the strong impression is left that this is what occurred in the past of us all, and still holds in many "savage" parts of the world.

> Man on an average is considerably taller, heavier, and stronger than woman, with squarer shoulders and more plainly pronounced muscles . . .
>   Man is more courageous, pugnacious and energetic than woman, and has a more inventive genius. His brain is absolutely larger, but whether or not proportionately to his larger body, has not, I believe, been fully ascertained. In woman the face is rounder; the jaws and the base of the skull smaller; the outlines of the body rounder, in parts more prominent; and her pelvis is broader than in man; but this latter character may perhaps be considered rather as a primary than a secondary sexual character. She comes to maturity at an earlier age than man.
>
> (Darwin 1871, 2: 316–17)

The discerning reader will have inferred already that women are truly juvenile males, and, for the not-so-discerning reader, Darwin himself helpfully drew the inference. "Male and female children resemble each other closely, like the young of so many other animals in which the adult sexes differ widely; they likewise resemble the mature female much more closely than the mature male" (p. 317).

How do we account for all of this?

> [L]ooking far enough back in the stream of time, and judging from the social habits of man as he now exists, the most probable view is that he aboriginally lived in small communities, each with a single wife, or if powerful with several, whom he jealously guarded against all other men. Or he may not have been a social animal, and yet have lived with several wives, like the gorilla; for all the natives "agree that but one adult male is seen in a band; when the young male grows up, a contest takes place for mastery, and the strongest, by killing and driving out the others, establishes himself as the head of the community." The younger males, being thus expelled and wandering about, would, when at last successful in finding a partner, prevent too close interbreeding within the limits of the same family.
>
> Although savages are now extremely licentious, and although communal marriages may formerly have largely prevailed, yet many tribes practise some form of marriage, but of a far more lax nature than that of civilised nations. Polygamy, as just stated, is almost universally followed by the leading men in every tribe.
>
> (pp. 362–3)

A number of other topics were also covered by Darwin, including infanticide.

> This practice is now very common throughout the world, and there is reason to believe that it prevailed much more extensively during former times. Barbarians find it difficult to support themselves and their children, and it is a simple plan to kill their infants . . . Wherever infanticide prevails the struggle for existence will be in so far less severe, and all the members of the tribe will have an almost equally good chance of rearing their few surviving children. In most cases a larger number of female than of male infants are destroyed, for it is obvious that the latter are of more value to the tribe, as they will, when grown up, aid in defending it, and can support themselves.
>
> (pp. 363–4)

The end of the whole discussion was to explain the differences between races. Why do we have white Europeans and black Africans and all of the others? Simply because of isolation and then the changing tastes in human beauty.

> We have seen that with the lowest savages the people of each tribe admire their own characteristic qualities,– the shape of the head and face, the squareness of the cheek-bones, the prominence or depression of the nose, the colour of the skin, the length of the hair on the head, the absence of hair on the face and body, or the presence of a great beard, and so forth. Hence these and other such points could hardly fail to be slowly and gradually exaggerated, from the more powerful and able men in each tribe, who would succeed in rearing the largest number of offspring, having selected during many generations for their wives the most strongly characterised and therefore most attractive women. For my own part I conclude that of all the causes which have led to the differences in external appearance between the races of man, and to a certain extent between man and the lower animals, sexual selection has been the most efficient.
>
> (p. 384)

Obviously much that Darwin was writing at this stage of the *Descent* was more a reflection of Victorian standards than disinterested science. His thinking was virtually guaranteed to appeal to his contemporaries – with the exception of J. S. Mill, whom Darwin took on with some gusto. "Now, when two men are put into competition, or a man with a woman, both possessed of every mental quality in equal perfection, save that one has higher energy, perseverance, and courage, the latter will generally become more eminent in every pursuit, and will gain the ascendancy." Apparently, Mill's views in *The Subjection of Women* that the "things in which man most excels woman are those which require most plodding, and long hammering at single thoughts" was confirmation of this position. Darwin's thinking, conversely, was virtually guaranteed to upset many today: "The *Origin* provided a mechanism for converting culturally entrenched ideas of female hierarchy into permanent, biologically determined, sexual hierarchy" (Erskine 1995, 118). The *Descent* apparently used that mechanism.

Before making an informed judgment on Darwin's thinking, trying to cut through the thickets of preconception, past and present, let us turn first to the history of thinking on human evolution –

paleoanthropology – from Darwin to the present. With more background context, we will be better able to return to the discussion of selection as it has shaped – and still shapes – humankind.

Human evolution

Huxley and company had good reasons for their enthusiasms. The post-*Origin* years were exciting times for the study of human evolution (Bowler 1986). By the time of the *Descent*, it was agreed by all that human remains had been discovered with the remains of animals now extinct, and even before the *Origin* (in 1856) specimens of a (supposedly) primitive kind of human had been unearthed in the Neander valley in Germany. (The name "Neanderthal" combines the proper name with the German word for "valley." In fact, in 1848 the skull of a human of this kind was uncovered in Gibraltar but was not recognized for what it was.) Although the Neanderthal skull is bigger than that of modern humans, general opinion was that these were primitive types, true "ape men." Naturally enough, more than a few were happy to identify them with certain living peoples in desolate parts of the earth. We find this sentiment continuing into the twentieth century. "Ferocious gorilla like living specimens of Neanderthal man are found not infrequently on the west coast of Ireland, and are easily recognized by the great upper lip, bridgeless nose, beetling brow with low growing hair, and wild and savage aspect" (Grant 1916, 108).

There was considerable debate as to whether Neanderthal man was a distinct species, the true missing link, between humans and brutes. Most, including Huxley (1863), who made it a key part of his discussion of human evolution, decided somewhat reluctantly that it was not. So the hunt continued for the crucial bridging fossils, many preferring Asia over Africa for the birthplace of humankind, less perhaps because of science than because it was thought degrading that white folk might have black ancestors. Far better that blacks are degenerated whites than that whites are advanced blacks. Finally, at the end of the century a Dutch doctor, Eugene Dubois, stationed in Indonesia, discovered a specimen from an earlier species of human-like form (known as "hominoid"). Popularly labeled "Java man" (reflecting the island of its discovery), it was catalogued by Dubois in a genus different from ours, *Pithecanthropus erectus*, although today we place it closer to us as *Homo erectus*. A common story – still a favorite with Creationists – that Dubois came to doubt the significance of his find is almost certainly not true.

Thanks to the greatest hoax in the history of science – Piltdown man – the search for human ancestors went badly off track in the first decades of the twentieth century. The south of England yielded specimens of a supposed ape-human, with a massive brain but primitive jaw. Just what the doctor ordered and a satisfyingly English answer to all of those nasty Neanderthals from the country across the channel with which Britain was now at war. We now know that Piltdown was a fake, half human and half orangutan, suitably stained and aged. To this day, no one is quite sure who was responsible, although incriminating chemicals were found in the property of one of the British Museum personnel, one Martin Hinton. (Given his whereabouts on key dates, it is unlikely that he worked alone.) The main scientific problem with the hoax was that it made the possibility of an African origin for humans seem even less likely, so that, when in the 1920s a human-like specimen with a very small skull was discovered in South Africa, it was dismissed as insignificant until at least the 1940s. Now "Taung baby," classified as *Australopithecus africanus*, is acknowledged to be the important find it is, and together with the many subsequent finds on that continent – a large proportion due to the energetic Leakey family – it has swung opinion right back to that of Darwin. The ancestors of *Homo sapiens* came from the Dark Continent, Africa (Johanson and Edey 1981).

What do we know today of human evolution? There are still large numbers of gaps, but the overall knowledge of our past is really quite complete (Wong 2003a). Much of this is based on fossil discoveries but, in the past three or four decades, molecular techniques have become increasingly important. Especially using the idea of a molecular clock, we can infer a lot about the past, including the very recent date of the break of humans from the shared branch with the higher apes. As is well known, it is even thought that we humans may be more closely related to the chimpanzees than the chimpanzees are to the gorillas. Tracing things right back, the earliest mammals probably appeared about 200 million years ago, although it was not until the dinosaurs disappeared 65 million years ago that we started to disperse and invade the whole world. Before that we were tiny, nocturnal beasties, trying to

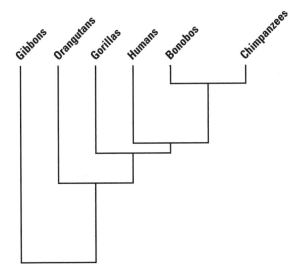

Figure 7.2 Human evolution and relation to the higher apes

keep out of the way. Primates came soon after, about 50 million years ago, and apes appeared about 20 million years ago. The molecular evidence suggests that humans emerged about 5 million years ago, and this is now starting to be backed by fossil discoveries from around that time or a little earlier (Figure 7.2).

The two most obvious differences between humans and the apes are the facts that we are bipedal and that we have big brains. No one, certainly not Darwin, thought that these are unconnected – as the hands are freed there is more scope for their intelligent use, and as we lose the power to run fast and climb and so forth, there is more need of intelligence and associated qualities. The discovery of Lucy, a comparatively full skeleton of *Australopithecus afarensis*, showed that by about 4 million years ago we were starting to get up on our hind legs; and discoveries of more recent fossils show that the brain was then on its way up to the present size – from around 400 cc to about 1,200 cc. (*Australopithecus afarensis* has a brain about the size of a chimpanzee's. This does not mean that the brain was that of a chimpanzee.)

Our genus, *Homo*, appeared between 1 and 2 million years ago – it is all a little bit arbitrary where you draw the line – and our species *Homo sapiens* about half a million years ago. There is much speculation about the origin of language, but it was certainly in place about 50,000 years ago, when human culture really started to explode in amount and sophistication (Mithen 1996). For a long time, there was uncertainty about the speaking abilities of Neanderthals, although now the pendulum has swung towards their being able to talk in some fashion. However, there is still a major debate about the relationship of modern humans to the Neanderthals (Ruse 2000; Wong 2003b). One group, the "multi-regional hypothesis" supporters, thinks that humans all evolved together – perhaps in different lines on different continents, but with enough crossbreeding that no group became extinct or broke off on its own. We are therefore descended from Neanderthals or at least carry many Neanderthal genes. The other group, the "Out of Africa hypothesis" supporters, thinks the ancestors of modern humans came out of Africa about 130,000 years ago, evolving into all of the different races we see today, and wiping out the Neanderthals along the way. The latter is probably the more favored of the two hypotheses by evolutionists, although one expects that there might have been some crossbreeding along the way and there do seem to be some hybrid specimens.

Notorious is the claim (made on the basis of cell parts known as "mitochondria") that all humans are descended from one female, inevitably nicknamed "Eve." Even if this were true, it would not mean that at one point there was one solitary female and she and she alone was the mother of modern humans (Ayala 1995). One expects that there was a group: if not there would have been very severe inbreeding, of which there is no evidence today. We are descended from other females as well as males. Probably, we are all descended from one or more other females and one or more other males.

Our story would be incomplete without mention of the most stunning scientific discoveries of the new millennium (Brown et al. 2004; Morwood et al. 2004). Back to Indonesia and another island, Flores, where there have been new discoveries of the most remarkable little being, human-like but small and with a small brain, three feet tall and with brain about 380 cc. Naturally nicknamed the "hobbit," the most remarkable thing is that this creature lived and flourished no more than 20,000 years ago and seems to have had sophisticated tool-use ability. Initially it was thought that this being was an immediate descendent of fairly recent members of the genus *Homo*, but now the suggestion is gaining ground that it might represent an independent line coming down straight from *Australopithecus*. Undoubtedly more specimens will be discovered and we shall know more about this mysterious little primate. Some doubters have suggested that the claims may all be overblown and the hobbit merely a deformed kind of human with a genetically shrunken brain; but this seems less and less likely.

Causes

Why did human evolution occur? Few today who work on these problems doubt the significance of selection, although literary critics and others have pointed out that, in traditional accounts of human evolution, there tends to be a fair measure of romanticism, not to mention fiction. We came down from the trees, unlike the chimps and the gorillas, and out onto the plains. Depending on your inclinations, either we were bold adventurers seeking new opportunities, or we were expelled from our jungle Eden by the cruel competitors who were stronger and luckier. Then we plucky little humans made the best of a bad situation, getting up on our hind legs since we no longer needed to climb and growing big brains to deal with life's challenges. And then, depending on how you look at it, either we all banded together on the principle enunciated by Benjamin Franklin on signing the Declaration of Independence: "Gentlemen, we must all hang together or assuredly we shall all hang separately," or we turned into the killer apes, which preyed on each other, destroying for the sheer pleasure of it all. And so the stories go.

Actually, there is a little more straw to make the bricks, even if there is still an awful lot of speculation. Start with the move to bipedalism. This was almost certainly linked to the decline of the jungles as a result of climatic changes, and the move to increasing areas of grassy plains. If you do not need to climb, then adopting a constant-standing position has many virtues. You can look out for predators like lions (also growing in number, given the increased numbers of herbivores on the plains). You can avoid getting too much sun, since your body is no longer fully exposed – although (as Darwin pointed out) your head is covered in hair for protection. And walking on two legs is much more energy efficient than knuckle walking, the way in which gorillas and chimpanzees get around.

When it comes to brains, the key factor is meat (Falk 2004). Today it may be possible to live comfortably as a vegetarian. Back then it was a different matter. The great apes will eat meat (monkeys and so forth) when they can, but generally they live on vegetable matter. Brains are expensive things to maintain. You need lots of high-quality, energy-making fodder, and the best source is meat. Carnivores almost inevitably have bigger brains than herbivores, although obviously, causally, some kind of feedback mechanism is at work here. Possibly our first bipedal ancestors were scavengers, simply primate jackals grabbing what they could from the spoils of large carnivores. With success came the opportunity to support more grey matter, and then the grey matter was used to the end of getting meat more efficiently, perhaps through cooperation, and so forth. So the increase in brain size came about, although as the hobbit suggests this was not always a one-way process. *Homo floresiensis*, as it is known, is not so surprising in one respect – namely, in being an island dweller. It is often the case that the inhabitants of islands will go to extremes with respect to size, becoming very much smaller than their mainland relatives, or very much bigger. A classic case of gigantism was the growth of huge birds like the moa on the islands of New Zealand (now all killed off by the Maori, who arrived in New Zealand from the islands of the Pacific). There were no mammals on New Zealand, so the birds were able to grow and take their place as the top predators. Conversely, on the island where the hobbit was found, there are the remains of very small elephants – no doubt shrunken through time because of selective forces triggered by a lack of foodstuffs. *Homo floresiensis*, who is thought to have hunted the elephants, may well have been reduced in parallel, although until more is known it is going to be hard to make any definitive judgments.

You may complain that much about the causes of human evolution is still speculative, and no one would argue strenuously against you. However, new techniques are revealing new information and making some hypotheses more probable than others. For instance, detailed study of teeth is very valuable for telling about diet, and study of tool use and of the bones of animals found with hominid remains tell much about the habits of our predecessors (Isaac 1983). Thinking that we may once have been scavengers is not simply a function of a Calvinistic mindcast – a self-deprecating belief that we must have been pathetic degenerates – but the conclusion of detailed examination of abilities to hunt and gather food and so forth.

Human sociobiology

We return now to the problem with which Darwin was wrestling in the *Descent* – namely, the extent to which humans have been shaped by selection, natural and sexual, and how this legacy lingers today. As we have just seen, biological opinion is that selection of one sort or another was very important in producing human beings. But has it left its mark? The general assumption by social scientists, happily endorsed by most people in the humanities – with many philosophers in the forefront – is that biology matters little if at all. Whatever may have happened in the past, humans through their culture have escaped their animal origins. Today we are different and, with respect to the effects of selection, autonomous. There are exceptions to this way of thinking. I am one. We shall encounter some others in subsequent chapters. We are in a minority, and, without seeming unduly paranoid, generally those of us (coming from areas like philosophy) who think that the genes, as chosen by natural selection, still play a significant role in human beliefs and actions, have been criticized strongly.

It is tempting at this point to paraphrase St. Paul: "If any man love not natural selection, let him be anathema, maranatha." Favoring selection as the determinant of human nature is the ultimate break with Western tribal beliefs. It is rather scary to think that we are really and truly at one with the organic world, and making this ultimate Darwinian commitment is to go against over 2,000 years of denial. One suspects that deep down in the minds of those who deny our animal nature lies a little residue of that form of Judeo-Christianity that sees humans as special, made in the image of God, and believes that this implies distinctive creative forces. For the most supposedly secular of thinkers, it is very hard to shake that Wittgensteinian prejudice about the irrelevance of evolution. Even those fond of evolution itself, for example the antipodean philosopher Kim Sterelny, strive to show that when it comes to *Homo australienthis* it is necessary to separate the hairiest human brute from its cousins in the mammalian world. What may have been is no more. "Our Pleistocene forebears did not have contemporary minds in a Pleistocene world; and we do not have essentially Pleistocene minds in a contemporary world" (Sterelny 2007). (The Pleistocene started 1.8 million years ago and ended 12,000 years ago.)

Most evolutionary biologists today, particularly those who work on the evolution of human social behaviour, human sociobiologists, would beg to differ. They think that Darwin was absolutely right in his approach – his approach, not necessarily his prejudices. Virtually no one wants to endorse the rather crude assumptions made by Darwin about male–female differences. Anyone who thinks that males are innately more intelligent than females should step inside undergraduate classrooms at universities today in North America. On average, the women outnumber the men by 60:40. But this is not to deny that biology has left its mark. Without at all endorsing the boorish behavior of fraternity jocks, there would be general agreement that male and female mating inclinations and practices really do differ, and that biology is at the root of these differences.

A teenager of my generation – I speak autobiographically – who greased back his hair and wore bell bottoms was as much an example of the power of culture as of the hidden control of the genes. Hence, biologists point out that it is often pointless to try to separate biology and culture. But culture is not the only factor. Males, and this includes human males, produce lots of sperm and potentially they can have many offspring. Just as potentially, if there is competition for mates, males might end up with no offspring at all. Females, and this includes human females, produce a limited number of eggs and equally their potentiality for reproduction is more limited. They will also have the burden of childrearing, so the nature of the sperm producer that fertilizes them is going to be of some concern. On the other hand, biology permitting, any female can get pregnant, and have some offspring. In humans, this leads to different inclinations and attitudes – notoriously, E. O. Wilson (1975) referred to women as "coy," although others (female primatologists) have suggested that "manipulative" might be a better term (Hrdy 1981) – and in turn this produces sexual selection, as is borne out by the fact that, as a species (Darwin was right) we are sexually dimorphic. On average males are bigger and stronger than females, although note that our sexual dimorphism is nothing like as extreme as that of sea lions, some species of deer, and primates like gorillas. Hence, the average biologist would agree with William James – and Darwin before him and Wilson after him – when he concludes that we are mildly polygamous.

Higgamous hoggamous,
Women are monogamous;
Hoggamous higgamous,
Men are polygamous!

This is not to deny that, in a rich society like ours, social factors can intervene and push us toward monogamy. Education, the ready availability of efficient contraception, determined pressure groups can and do make a difference. Although in a society like ours, where almost 50 percent of marriages end in divorce, it is hard really to think of us as truly monogamous. Moreover, with the fact that DNA testing suggests that up to 10 percent of paternities might not reflect social arrangements, together with the additional fact that many high-status men seek out repeated matings with younger partners – "trophy wives" – it would be a rash conclusion that, even in Western culture, sexual selection no longer operates.

In other spheres, likewise, there does seem to be evidence that selection and its effects – sometimes natural selection, sometimes sexual selection – are still to be found in human societies. A well-known study of parental abuse of children has shown, with much convincing detail, that adults in reconstituted relationships – particularly men in such relationships – are much more likely to abuse children of their mates than children to whom they are themselves genetically related. In other words, stepfathers show more violence toward the children than do biological fathers. This is not to say that most stepfathers are violent – they are not – but it is to say that stepchildren stand in more danger from a stepfather than from a biological father. In fact, in much more danger. "An American child living with one or more substitute parents in 1976 was . . . approximately 100 times as likely to be fatally abused as a child living with natural parents only" (Daly and Wilson 1988, 89). And what suggests that this is not merely cultural is the fact that these statistics hold for both the US and Canada. Violence – particularly homicide – is much more prevalent in the US than Canada, no doubt due to cultural factors like the ready availability of guns in the one country but not in the other. When it comes to step-parental abuse, however, the comparative differences vanish.

A major reason why biologists – human sociobiologists, or as they are often called today "evolutionary psychologists" – think that biology might be important in the violence case, as in the polygamy case, is that this is an instance of a phenomenon already well recorded in the animal world: a phenomenon with a ready selective explanation. To take but three non-human examples from many: in the langur, lemming, and lion worlds, when males move in and take over a pride or any female, the first act is to kill off all of the existing young offspring. This brings the female back into heat and ensures that the offspring of the new male will get full attention. Note that no one is saying that male humans, any more than male langurs or male lemmings or male lions, are consciously thinking in these terms. One truly would be insane if one killed a child because one had done some simple population-genetical calculations. Nor is anyone saying that, in today's world, such behavior is at all likely to have a biological payoff. Prison is more likely. But it is to say that the way in which humans behave today, even in so-called civilized societies, might show traces of that very Pleistocene ancestry about which Sterelny is so dismissive.

What about biological parental infanticide, the topic that so fascinated Darwin? Primatologist Sarah Blaffer Hrdy (1999) argues that not only is infanticide common in human societies but that it is strongly controlled by biological factors. She agrees with Darwin that it is often female infants that are killed, but she wants to take the argument one step further and claims that the sex of the child being killed is a function of the status and situation of the parents. Darwin was right that where families need sons – for instance, to protect the group or to carry on the business – then infanticide will tend to preserve males and destroy females. But where families have lower status, then at best sons tend to be left to fend for themselves and it is daughters who are cherished – if only because, through matches to high-status sons, the whole family benefits. Hrdy points out that this is an exemplar of a well-known theorem (backed by much empirical evidence) in the biological world that suggests that high-status mothers will tend to produce males and low-status mothers to produce females. In the human world, this may be done through intention and culture. In the biological world, it is physiology that does the trick. But in both

Humans

cases the result is the same. If it is possible to produce high-quality sons, then take this strategy: they may well score highly in the struggle for existence. Otherwise, produce daughters: they will invariably give some return, however meager, and if they mate with high-quality males then the returns may be greater.

Darwin is surely right. Biology does matter. Yet – as we shall see later – in many respects, Sterelny and fellow thinkers are also surely right. Humans are more than inherited biology. Let us leave it at that, without estimating who is more right than the other. In following chapters, we shall be returning to some of these issues, as we wrestle with the importance of biology for human knowledge and morality.

Progress

But before we can do this, we must turn to a major philosophical issue that lies always behind discussions about the evolution of humans (Ruse 1996). Was our appearance somehow necessary, or at least to be expected – if not exactly as we are, then at least with consciousness? And is it correct to judge conscious beings like us humans as superior to all others? In other words, is evolution progressive and have we won? For all of the diversions and regressions, is not the story of evolution one of upward climb, somehow something that was bound to be? And did humans come top? As the paleontologist G. G. Simpson used to say, if others think that they have won then why don't they speak up and say so?

Historically speaking, every evolutionist before Charles Darwin thought that evolution was progressive. Indeed, that was a major reason – at times it seems the only reason – why people endorsed evolution. They started with beliefs in progress in the social and cultural world, and then read this into the biological world, getting evolution as a result. Generally, they then read this back into the cultural and social world as confirmation of their original beliefs! Erasmus Darwin, Charles's grandfather, was a paradigm. As he himself put it, life was a progress from monarch (butterfly) to monarch (king).

Organic Life beneath the shoreless waves
Was born and nurs'd in Ocean's pearly caves;
First forms minute, unseen by spheric glass,

Move on the mud, or pierce the watery mass;
These, as successive generations bloom,
New powers acquire, and larger limbs assume;
Whence countless groups of vegetation spring,
And breathing realms of fin, and feet, and wing.

Thus the tall Oak, the giant of the wood,
Which bears Britannia's thunders on the flood;
The Whale, unmeasured monster of the main,
The lordly Lion, monarch of the plain,
The Eagle soaring in the realms of air,
Whose eye undazzled drinks the solar glare,
Imperious man, who rules the bestial crowd,
Of language, reason, and reflection proud,
With brow erect who scorns this earthy sod,
And styles himself the image of his God;
Arose from rudiments of form and sense,
An embryon point, or microscopic ens!

(Darwin 1803, 1: ll. 295–314)

Erasmus Darwin would complement these ideas with other verses celebrating the triumphs of his fellow men of business and industry:

So with strong arm immortal BRINDLEY leads
His long canals, and parts the velvet meads.

(Darwin 1791, pt. 3: 349–50)

He drew an explicit analogy between the upward path of culture and that of biology, the two notions really being but one. The idea of organic progressive evolution "is analogous to the improving excellence observable in every part of the creation; . . . such as the progressive increase of the wisdom and happiness of its inhabitants" (Darwin 1794–6, 1: 509).

At the time of Charles Darwin these sorts of sentiments still were going strong. Nature is progressive, and this applies to organisms and knowledge indifferently. That which comes last is better in some sense. The English man of science and letters Herbert Spencer was pushing exactly this kind of line. An ardent Lamarckian – the chief cause of change is the inheritance of acquired characters – Spencer saw organic evolution as being but one facet of the overall upward progress that characterizes the whole world process: from the undifferentiated to the differentiated, or in his words from the homogeneous to the heterogeneous:

> Now we propose in the first place to show, that this law of organic progress is the law of all progress. Whether it be in the development of the Earth, in the development of Life upon its surface, in the development of Society, of Government, of Manufactures, of Commerce, of Language, Literature, Science, Art, this same evolution of the simple into the complex, through successive differentiations, holds throughout. From the earliest traceable cosmical changes down to the latest results of civilization, we shall find that the transformation of the homogeneous into the heterogeneous, is that in which Progress essentially consists.
>
> (Spencer 1857, 246)

Nothing escapes this law. Humans are more complex or heterogeneous than other animals; Europeans are more complex or heterogeneous than savages; and (hardly a surprise) the English language is more complex or heterogeneous than the languages of other speakers.

Darwinism

Surely Darwin smashed this kind of thinking to smithereens? After all, the whole point of natural selection is that it is relative to the situation. It is no tautology, for that which wins in one case is not necessarily that which wins in another, different, case. Is it better to be red or green? It depends. If the predators can spot red over green, better to be green. If green over red, better to be red. In the jungle, better to be with four limbs all suited for climbing. If the jungle recedes, then better to be on two legs rather than four, even though climbing ability is lost. But, as is so often true, it is important here to separate what may be conceptually the case with what was actually the case. Historically, do not rush too quickly. Darwin was certainly keen to avoid suggestions that nature naturally and necessarily leads upward to perfection. As we saw earlier, this smacks too much of a kind of necessity in nature – in fact, it smacks too much of the thinking of German idealists like the philosopher G. W. F. Hegel (1817), who subscribed to the nature philosophy, *Naturphilosophie*, that English biologists like Darwin suspected and wanted to keep at arm's length. There is too much of an odor of a kind of philosophical or theological necessity that was alien to the empiricism of the British.

Having said this, however, Darwin was as keen on progress as the next person – and why not, since he had benefited so much from England's success in industry and technology and the like? He was as happy as his grandfather to jump from culture to biology. For all that he wanted to keep values out of the *Origin*, progressivist sentiments do come through. Consider the penultimate paragraph:

> As all the living forms of life are the lineal descendants of those which lived long before the Silurian epoch, we may feel certain that the ordinary succession by generation has never once been broken, and that no cataclysm has desolated the whole world. Hence we may look with some confidence to a secure future of equally inappreciable length. And as natural selection works solely by and for the good of each being, all corporeal and mental endowments will tend to progress towards perfection.
>
> (Darwin 1859, 489)

The Descent of Man is quite open:

> In the class of mammals the steps are not difficult to conceive which led from the ancient Monotremata to the ancient marsupials; and from these to the early progenitors of the placental mammals. We may thus ascend to the Lemuridae; and the interval is not very wide from these to the Simiadae. The Simiadae then branched off into two great stems, the New World and Old World monkeys; and from the latter, at a remote period, Man, the wonder and glory of the Universe, proceeded.
>
> (Darwin 1871, 1: 213)

How can selection possibly work to produce humans, if not by some kind of mystical necessity, then at least as the expected outcome? Although he was never terribly explicit, Darwin seemed to think that something akin to what today we would label an "arms race" might do the trick. Evolving lines of organisms compete against each other, and improvement occurs – the prey gets faster and hence the predator gets faster and so forth. Remember: "The inhabitants of each successive period of the world's history have beaten their predecessors in the race for life, and are, in so far, higher in the scale of nature; and this may account for that vague yet ill-defined sentiment, felt by many palæontologists, that organisation on the whole has progressed" (Darwin 1859, 345). But can this account for intelligence? Apparently so.

> If we take as the standard of high organisation, the amount of differentiation and specialization of the several organs in each being when adult (and this will include the advancement of the brain for intellectual purposes), natural selection clearly leads towards this standard: for all physiologists admit that the specialization of organs, inasmuch as in this state they perform their functions better, is an advantage to each being; and hence the accumulation of variations tending towards specialisation is within the scope of natural selection.
>
> (Darwin 1959, 222; this passage was added
> in the third edition, 1861)

If Darwin did not grasp fully the implications of his theory, did subsequent events and discoveries do the job for him? The twentieth century saw the coming of Mendelism. Is not the non-directedness of mutation another nail in the coffin of progress? How can you have upward direction if the building blocks go every which way? Again, history trumps concepts! Mendelism proved no barrier at all to thoughts of progress. Evolutionists – Darwinians – went right on believing in upward progress and that humans have won the prize. Arms races continued to figure high in people's thinking about how selection can lead to improvement. In a little book, *The Individual in the Animal Kingdom*, written before the First World War, Julian Huxley likened biological evolution to the competition between nations in preparation for war. Germany and Britain were competing on the sea, leading Huxley to write: "The leaden plum-puddings were not unfairly matched against the wooden walls of Nelson's day." He then added that in his day "though our guns can hurl a third of a ton of sharp-nosed steel with dynamite entrails for a dozen miles, yet they are confronted with twelve-inch armor of backed and hardened steel, water-tight compartments, and targets moving thirty miles an hour. Each advance in attack has brought forth, as if by magic, a corresponding advance in defence" (Huxley 1912, 115–16).

Explicitly, Huxley likened this to the organic world, for:

> if one species happens to vary in the direction of greater independence, the inter-related equilibrium is upset, and cannot be restored until a number

of competing species have either given way to the increased pressure and become extinct, or else have answered pressure with pressure, and kept the first species in its place by themselves too discovering means of adding to their independence.

(Huxley 1912, 116)

And so finally: "it comes to pass that the continuous change which is passing through the organic world appears as a succession of phases of equilibrium, each one on a higher average plane of independence than the one before, and each inevitably calling up and giving place to one still higher." This argument was repeated in *Animal Biology* (1927), a classic textbook coauthored with J. B. S. Haldane, and in Huxley's own masterwork, *Evolution: The Modern Synthesis* (1942).

Progress today?

Where do we stand today on the question of biological progress? Some of today's most distinguished Darwinians are ardent for the topic. Edward O. Wilson writes:

the overall average across the history of life has moved from the simple and few to the more complex and numerous. During the past billion years, animals as a whole evolved upward in body size, feeding and defensive techniques, brain and behavioral complexity, social organization, and precision of environmental control – in each case farther from the nonliving state than their simpler antecedents did.

(Wilson 1992, 187)

He adds: "Progress, then, is a property of the evolution of life as a whole by almost any conceivable intuitive standard, including the acquisition of goals and intentions in the behavior of animals." At the general level, Wilson does not really try to explain progress. He more or less takes it as a given. In the social world, however, he has speculated on causes, particularly on the evolution of humankind, a species that he thinks has reversed the downward trend in sophistication of social interactions since the Hymenoptera (the ants, the bees, and the wasps) first appeared in the world. But we do seem to be rather a one off, since clearly what we do is a function of our large brains,

Humans

and hence one can hardly say that what happened with us is a model for what happens throughout the history of life.

Richard Dawkins tries to give a more general account of progress, continuing and refining the arms race approach. Like Wilson he is a committed progressionist: "Directionalist common sense surely wins on the very long time scale: once there was only blue-green slime and now there are sharp-eyed metazoa" (Dawkins and Krebs 1979, 508). Adaptation is at the heart of his definition of progress: "A tendency for lineages to improve cumulatively their adaptive fit to their particular way of life, by increasing the numbers of features which combine together in adaptive complexes." There is no direct mention here of humans, but he clearly thinks that humans come out well on this definition. The point is that for Dawkins, unlike Wilson, humans are not unique in this sense. What makes for progress to humans was what made for progress to other organisms that came out higher than their ancestors. We are the winners, but we succeeded by the same means as others.

Spelling out his position in his great popular overview of modern evolutionary thinking, *The Blind Watchmaker*, Dawkins uses the arms race concept, suggesting that not only do such races produce advanced or better adaptations: they lead ultimately to progress, and eventually progress leads to humans, simply because, when all is said and done, brains are the very best of all adaptations. After all, the history of arms races in the last century shows that increasingly military strategy depends less on sheer brute force and more on sophisticated weaponry using high-tech electronic equipment. These are analogous to the development of organisms' on-board computers, better known as brains. Dawkins refers to Harry Jerison's notion (1973) of an "Encephalization Quotient" (EQ), a kind of universal animal IQ: it takes brain size and subtracts the gray matter needed simply to get the body functioning – whales necessarily have bigger brains than shrews, because they have bigger bodies – and counts what is left when you take off this body-functioning portion. Measured thus, humans come way out on top, leading Dawkins (1986, 189) to reflect: "The fact that humans have an EQ of 7 and hippos an EQ of 0.3 may not literally mean that humans are 23 times as clever as hippos!" But, he concludes, it does tell us "something."

Elsewhere, Dawkins has tied in his thinking about progress with the notion of the "evolution of evolvability." Sometimes, you just get evolutionary breakthroughs – like the eukaryotic cell – that have more potential, and evolution has made a jump to a new dimension. "There really is a good possibility that major innovations in embryological technique open up new vistas of evolutionary possibility and that these constitute genuinely progressive improvements" (Dawkins 1989; cf. Maynard Smith and Szathmáry 1995).

> The origin of the chromosome, of the bounded cell, of organized meiosis, diploidy and sex, of the eucaryotic cell, of multicellularity, of gastrulation, of molluscan torsion, of segmentation – each of these may have constituted a watershed event in the history of life. Not just in the normal Darwinian sense of assisting individuals to survive and reproduce, but watershed in the sense of boosting evolution itself in ways that seem entitled to the label progressive. It may well be that after, say, the invention of multicellularity, or the invention of metamerism, evolution was never the same again. In this sense, there may be a one-way ratchet of progressive innovation in evolution.
>
> (Dawkins 1997, 1019–1020)

Dawkins has always made brilliant use of metaphor – selfish gene, blind watchmaker, mount improbable (to echo the titles of some of his books) – and metaphor is much involved in the thinking about progress. In The Blind Watchmaker, the metaphor of bigger and bigger on-board computers (aka brains) plays a vital role, as it has elsewhere.

> Computer evolution in human technology is enormously rapid and unmistakably progressive. It comes about through at least partly a kind of hardware/software coevolution. Advances in hardware are in step with advances in software. There is also software/software coevolution. Advances in software make possible not only improvements in short-term computational efficiency – although they certainly do that – they also make possible further advances in the evolution of the software. So the first point is just the sheer adaptedness of the advances of software make for efficient computing. The second point is the progressive thing. The advances of software open the door – again I wouldn't mind using the word "floodgates" in some instances – open the floodgates to further advances in software.
>
> (Ruse 1996, 469; from a presentation given in Melbu, Norway, in 1989)

Evolution is cumulative, for it has "the power to build new progress on the shoulders of earlier generations of progress." And brains, especially the biggest and best brains, are right there at the heart, or (perhaps we should say) end: "I was trying to suggest by my analogy

Humans

with software/software coevolution, in brain evolution that these may have been advances that will come under the heading of the evolution of evolvability in [the] evolution of intelligence" (Ruse 1996, 469).

Recently, the Cambridge paleontologist Simon Conway-Morris (2003) has tried another tack, making a vigorous, Darwin-inspired attempt to refurbish a sense of progress, based on a selection-driven concept of adaptation. Conway-Morris's starting position is that only certain areas of potential morphological space are going to be capable of supporting functional life, and to this he adds the assumption that selection is forever pressing organisms to look for such potential, functional spaces. Hence, if such spaces exist, sooner or later they will be occupied – probably sooner rather than later, and probably many times. Conway-Morris draws attention to the way in which life's history shows an enormous number of instances of convergence – instances where the same adaptive morphological space has been occupied again and again. The most dramatic perhaps is the case of saber-toothed-tiger-like organisms, where the North American placental mammals (real cats) were matched item for item by South American marsupials (thylacosmilids). Clearly there existed a niche for organisms that were predators, with cat-like abilities and shearing/stabbing weapons, and natural selection found more than one way to enter it. Indeed, it has been suggested, long before the mammals, the dinosaurs might also have found this niche (Figure 7.3).

Conway-Morris argues that this pattern happens over and over again, showing that the historical course of nature is not random but strongly selection-constrained along certain pathways and to certain destinations. From this, he concludes that movement up the order of nature, the chain of being, is bound to happen, and eventually some kind of intelligent being (what has been termed a "humanoid") is bound to emerge. We know from our own existence that a kind of cultural adaptive niche exists – a niche where intelligence and social abilities are the defining features. Moreover, we know that this niche is one to which other organisms have (with greater or lesser success) aspired. We know of the kinds of features (like eyes and ears and other sensory mechanisms) that have been used by organisms to enter new niches; we know that brains have increased as selection presses organisms to ever new and empty niches; and we know that, with this improved hardware, have come better patterns of behavior and so forth (more sophisticated software). Could this not all add up to something?

Figure 7.3 Saber-toothed mammals: marsupial thylacosmilid (*top*) and placental cat (*bottom*)

If brains can get big independently and provide a neural machine capable of handling a highly complex environment, then perhaps there are other parallels, other convergences that drive some groups towards complexity. Could the story of sensory perception be one clue that, given time, evolution will inevitably lead not only to the emergence of such properties as intelligence, but also to other complexities, such as, say, agriculture and culture, that we tend to regard as the prerogative of the human? We may be unique, but paradoxically those properties that define our uniqueness can still be inherent in the evolutionary process. In other words, if we humans had not evolved then something more-or-less identical would have emerged sooner or later.

(Conway-Morris 2003, 196)

Problems

With all of this enthusiasm for progress, you might think that progress is the end of matters. Be not so quick. There have always been dissenters, starting with Thomas Henry Huxley, who swung from enthusiasm to doubt. He did not want to deny that more and more sophisticated organisms are produced by evolution, but toward the end of his life (he died in 1895) he did want to query whether this was always progress.

> Man, the animal, in fact, has worked his way to the headship of the sentient world, and has become the superb animal which he is, in virtue of his success in the struggle for existence. The conditions having been of a certain order, man's organization has adjusted itself to them better than that of his competitors in the cosmic strife. In the case of mankind, the self-assertion, the unscrupulous seizing upon all that can be grasped, the tenacious holding of all that can be kept, which constitute the essence of the struggle for existence, have answered. For his successful progress, throughout the savage state, man has been largely indebted to those qualities which he shares with the ape and the tiger; his exceptional physical organization; his cunning, his sociability, his curiosity, and his imitativeness; his ruthless and ferocious destructiveness when his anger is roused by opposition.
>
> But, in proportion as men have passed from anarchy to social organization, and in proportion as civilization has grown in worth, these deeply ingrained serviceable qualities have become defects. After the manner of successful persons, civilized man would gladly kick down the ladder by which he has climbed. He would be only too pleased to see "the ape and tiger die." But they decline to suit his convenience; and the unwelcome intrusion of these boon companions of his hot youth into the ranged existence of civil life adds pains and griefs, innumerable and immeasurably great, to those which the cosmic process necessarily brings on the mere animal.
>
> (Huxley 1989, 51–2)

More recently, Stephen Jay Gould was a strong critic of the notion of progress, speaking of it as "a noxious, culturally embedded, untestable, nonoperational, intractable idea that must be replaced if we wish to understand the patterns of history" (Gould 1988, 319). It is a delusion engendered by our refusal to accept our insignificance when faced with the immensity of time (Gould 1996). Making reference to the asteroid that hit the earth some 65 million years ago and that (we think) wiped out the dinosaurs, Gould wrote:

Since dinosaurs were not moving toward markedly larger brains, and since such a prospect may lie outside the capabilities of reptilian design . . . , we must assume that consciousness would not have evolved on our planet if a cosmic catastrophe had not claimed the dinosaurs as victims. In an entirely literal sense, we owe our existence, as large and reasoning mammals, to our lucky stars.

(Gould 1989, 318)

There are two questions here, one empirical and the other more philosophical. At the empirical level, the issue is whether the mechanisms proposed will do the trick. Can we reasonably expect that selection will produce complex organisms, and most particularly humans or human-like creatures? Some are not so sure. There are facts both for and against the efficacy of arms races, for instance. Over the long haul, the fossil record does not always show that competition leads to advance (Bakker 1983). There is evidence that predator–prey interactions do not always lead to great speed, for instance. (There is, to be fair, evidence that shells get thicker and the boring apparatuses of attackers get more sophisticated (Vermeij 1987).) But even if arms races do work, will they necessarily lead to intelligence? We think so, but we would! It happened once but would it be expected to happen on a regular basis? We have seen that intelligence does not come cheaply. Brains need lots of protein. *Homo floresiensis* suggests that sometimes you might be better off not going for the big-brain option. The point is that natural selection apparently can lead to brains, but whether it would always do so or even is likely to do so is another matter.

Similar doubts surround the Conway-Morris proposal. We all have a notion of successive niches – sea, land, air, and culture – but this is a bit flawed. My suspicion is that, apart from the troublesome assumption that water, land, air, and culture make for a simple progression – Why is the land necessarily superior to the water? Are dogs superior to whales? – there is the assumption of pre-existing niches. But do these make sense? In his *The Extended Phenotype*, Dawkins shows that organisms are involved in their surroundings, their niches, and that often if not always it is difficult to distinguish the two. You have a beaver and you have its lodge. Where does the beaver end and the lodge begin, where does the animal end and its niche begin? From a Darwinian perspective, these are not easy or straightforward questions. If the lodge is as important to the beaver's wellbeing as the tail, and if the beaver did as much in creating the lodge as in creating the tail, why then make an ontological separation between the two? In some sense, organisms create their own niches. And if this is so, then what price culture? We have created it, but would it ever occur again? Or would it occur in such a way that we might not want to talk of intelligence or whatever? Conway-Morris surely scores a good point in noting that organisms do seem to converge on the same spots, suggesting that niche creation is not entirely arbitrary. But clearly more work is needed on this topic.

Defining progress

Turn now to the more philosophical issue (Ruse 1993). There are two matters here. First, can we really give a definition of progress? Many would appeal to complexity as a help here. Richard Dawkins talks in terms of "adaptive fit," and he thinks that we are building up "adaptive complexes." But what is complexity? Starting with ideas in information theory, Dawkins thinks that more complex organisms would require physically longer descriptions than less complex organisms.

> We have an intuitive sense that a lobster, say, is more complex (more "advanced", some might even say more "highly evolved") than another animal, perhaps a millipede. Can we measure something in order to confirm or deny our intuition? Without literally turning it into bits, we can make an approximate estimate of the information contents of the two bodies as follows. Imagine writing the book describing the lobster. Now write another book describing the millipede down to the same level of detail. Divide the word-count in the one book by the word-count in the other, and you will have an approximate estimate of the relative information content of lobster and millipede. It is important to specify that both books describe their respective animals "down to the same level of detail". Obviously, if we describe the millipede down to cellular detail, but stick to gross anatomical features in the case of the lobster, the millipede would come out ahead.
>
> But if we do the test fairly, I'll bet the lobster book would come out longer than the millipede book.
>
> (Dawkins 2003, 100)

But even if this works, does it really yield a notion of progress? Paleontologist Dan McShea (1991) has pointed out that the backbone of the whale is significantly simpler by any measure than that of

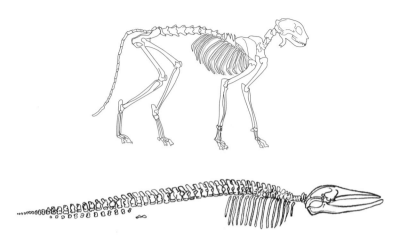

Figure 7.4 Comparison of the skeletons of the tiger (*top*) and the whale (*bottom*)

a land carnivore, like a tiger (Figure 7.4). But would one want to say that the tiger is superior to the whale? Perhaps you might argue that it is not just a question of one feature but of all features. The whale has other adaptations not possessed by the tiger. For instance, it is able to withstand the cold of the ocean and it is able to dive much deeper and stay down for extended periods of time. Overall the whale is not worse off than the tiger. But granting that brains are very complex things, why does the human beat the whale, taken overall? We have complex adaptations for what we do, and the whale has complex adaptations for what it does, and that is the end of the matter. Of course, you might want to say – as Dawkins certainly does in *The Blind Watchmaker* – that the human brain is better than the brain of anything else. But now we are starting to shift the definition, away from complexity as such to brain complexity. We can certainly make humans the top of the progress chart if we define progress features in terms of what we humans have, but that does rather defeat the whole exercise.

Note that I am not saying that it is impossible to define progress. You can always do that if you want; although whether it gives you what you need – what your human intuitions tell you are the marks of progress – is another matter. For instance, for a while the idea that the DNA would yield a definition of progress was a favorite (Maynard

Smith 1988). But then it turned out that humans have less DNA in their cells than some amphibians. As the work on the human genome project has shown very well, it is not a question of how many genes one has – how much DNA is there, or even how much functioning DNA is there – but rather of what one does with it. Humans presumably do more than amphibians. Other measures have been tried: for instance, the length of time that a species lasts. But even though there is some evidence that species today do (or before human activity, did) last longer than in days of yore, again it is not easy to see exactly how this gives humans the edge. Some slow-developing and slow-changing organisms, like horseshoe crabs, may well turn out to be winners.

Really, one reverts here to the point that Whewell made about classification. No one is saying that it is logically impossible to find some characteristic or set of characteristics that put humans on the top rung. But if this is to be an objective measure and not just redescription and self-promotion, these characteristics cannot be defined straight off in terms of human-like features. They must be independent if the measure is to have any pretensions to objectivity. However, for all the searching, so far it is not obvious that such a set of characteristics has been found. And this leads to the second point. Even if one does find such a set of characteristics, why should they be defined or described in terms of value? Suppose DNA content in cells had correlated with complexity. Why should it be better to have more DNA than less? It is certainly not obvious that complexity is going to be better – not in any biological sense, for sure. In the memorable words of paleontologist Jack Sepkoski: "I see intelligence as just one of a variety of adaptations among tetrapods for survival. Running fast in a herd while being as dumb as shit, I think, is a very good adaptation for survival" (Ruse 1996, 486).

There is a basic problem here. Modern science, as understood by us – as Darwin tried to practice it in the *Origin* (not always successfully, as is shown by his wrestling with the notion of progress) – is designed to be value free (Nagel 1961). At least, absolute-value free. You can have relative values like "whiter than" or "weaker than," but you cannot have absolute values like "Christianity is the best religion" or "Europeans are superior to Africans." You yourself might think that Christianity is the best religion and (although you may not dare to say so) that Europeans are the top humans around. But this is not something you get from science, which is designed explicitly to be value free. It is supposed to give a disinterested picture of objective reality. Notions like "feminist science" or "Jewish science" are oxymorons like "weapons for peace" or "famous Canadians." So if you want to say something like: "Evolution has produced a superior sort of being, namely humans," you are at perfect liberty to do so – frankly, speaking as one human to another, I would think you a little silly if you did not – but it is something that you are reading into nature rather than deriving from it. Believe in biological progress to beings of absolute worth, but do not kid yourself that Darwinism is telling you this.

A final word

If you sense that I am severely conflicted on the subject of progress, then you sense correctly. I shall have more to say about progress when I get to the discussion of religion. For now, I have to say that I find the notion of biological progress attractive but I cannot see any basis for it. This leads me to suspect that I may be the victim of some kind of intellectual confidence trick. Am I caught in a Cartesian conundrum? Descartes said: "I think therefore I am." If I doubt this, then I reaffirm its truth. Is there progress? Remember what Simpson said about the need of other animals to speak up if they think that humans did not win the race. If you doubt that there is progress, then you affirm that there is, because you are capable of discussing the issue and to do this you needed the powers of thought. Combine this fact with the other fact that in order to discuss the issue at all we must be alive, which means we must be at the end point of evolution (so far), and it is little wonder that progress is such a seductive idea. Unfortunately, not all who seduce are entirely trustworthy.

8

Knowledge

If, indeed, we humans were not created miraculously on the sixth day, but are the end products of a long, slow process of evolution, then one would expect this to have some implications for the big questions of philosophy: "What can I know?" and "What should I do?" Charles Darwin himself certainly thought it did, and so have others in the years since the *Origin*. In this and the next chapter we are concerned with the thinking of Darwin and his successors first about the theory of knowledge (epistemology) and then the theory of morality (ethics). As I pointed out in the Preface, Darwin was first and foremost a scientist rather than a philosopher. He does have interesting things to say about philosophical issues, and offers quite a detailed discussion of ethics, but overall his significance lies in the importance of his ideas for later thinkers. Hence, particularly in this chapter, I shall locate him in the tradition rather than highlight him.

Pragmatism

The American philosophy known as "pragmatism" defines knowledge in terms of results. Here is William James:

> The pragmatic method in such cases is to try to interpret each notion by tracing its respective practical consequences. What difference would it practically make to any one if this notion rather than that notion were true? If no practical difference whatever can be traced, then the alternatives mean practically the same thing, and all dispute is idle. Whenever a dispute is serious, we ought to be able to show some practical difference that must follow from one side or the other's being right.
>
> (James 1967, 322)

Given its down-to-earth attitude toward knowledge – what works is what counts – you might think that this would fit nicely with Darwinian thinking, and you will not be surprised to learn that the pragmatists were enthusiastic about the importance and relevance of evolution for their philosophy.

But there are differences between the responses of the pragmatists. John Dewey, in a famous lecture entitled "Darwinism and Philosophy" (1909), really committed himself only to the relativity of inquiry. Thanks to Darwin we know that all is in flux and what works at one time might not work at another. So it is with intellectual problems and answers. They change over time. "The influence of Darwin upon philosophy resides in his having conquered the phenomena of life for the principle of transition, and thereby freed the new logic for application to mind and morals and life" (p. 8). Incidentally, this is basically the position of today's great champion of Dewey, Richard Rorty. He speaks in high terms of Darwin:

> My hunch is that, just as we see philosophy from 1630 to 1800 ("Descartes to Kant") as an attempt to come to terms with corpuscularian mechanics, future historians will see philosophy from 1860 to (at least) 2060 as an attempt to come to terms with "the biological conception of the psyche" that Dewey found in James, and which James found developed in response to Huxley and Darwin.
>
> (Rorty 1995, 71)

But the exploration of this hunch seems to require little actual grappling with Darwin's thinking, as such, just the general recognition of change and of our animal natures. We will learn in the next chapter that Rorty is not inclined to give biology much of a role when it comes to morality.

Going back to Charles Sanders Peirce, by common agreement the greatest of all the pragmatist philosophers, we find a much deeper commitment to evolution as a mode of philosophical inquiry. For him, the world is in a transitional state and this reflects into understanding. The metaphysics of his "pragmaticism" (as he liked to call his position) was a form of psycho-physical realism, where body and mind intertwine and in some way point to an omega of total rationality, where chance will have vanished, and we shall have achieved a state of "evolutionary love." Speaking of the need for a "Cosmogonic Philosophy" Peirce wrote:

It would suppose that in the beginning – infinitely remote – there was a chaos of unpersonalized feeling, which being without connection or regularity would properly be without existence. This feeling, sporting here and there in pure arbitrariness, would have started the germ of a generalizing tendency. Its other sportings would be evanescent, but this would have a growing virtue. Thus, the tendency to habit would be started; and from this, with the other principles of evolution, all the regularities of the universe would be evolved. At any time, however, an element of pure chance survives and will remain until the world becomes an absolutely perfect, rational, and symmetrical system, in which mind is at last crystallized in the infinitely distant future.

(Peirce 1958, 158–9)

One hardly needs to say that this is a long way from Charles Darwin, and to be frank it seems to owe a lot more to German idealism, particularly the thought of Hegel, than to British natural history. We know that Peirce was much influenced by early nineteenth-century, continental thought – as well as by the neo-Kantian account of the history of science by none other than William Whewell – and we have seen already that, although Peirce was able to offer an astute analysis of the statistical nature of Darwinism, he had little sympathy for the mechanisms of Darwinism. Truly he loathed what he saw were the vile social implications of Darwinism – the "greed-philosophy" – but also he had been seduced by the non-evolutionary thinking of Louis Agassiz (remember, now at Harvard). In the post-*Origin* years, it was easy enough to ignore the non-evolutionism and yet still to embrace the upward thrust of the *Naturphilosoph* view of life's history. Although Peirce became a transformist, he always inclined to jumps, saltations, as well as to a dash of Lamarckism. Natural selection came a poor third.

The person who really took on not just evolution but Darwinism was William James. His essay "Great men, great thoughts, and the environment" (published in 1880), went all of the way. "A remarkable parallel, which I think has never been noticed, obtains between the facts of social evolution on the one hand, and of zoölogical evolution as expounded by Mr. Darwin on the other" (James 1956, 261). He was no enthusiast for Lamarckism and saw clearly that Darwinism has two essential components: the new variations which, although not uncaused, do not appear to be directed towards need; and the effects of the environment leading to selection, the results of which in turn feed back into the environment.

The causes of production of great men lie in a sphere wholly inaccessible to the social philosopher. He must simply accept geniuses as data, just as Darwin accepts his spontaneous variations. For him, as for Darwin, the only problem is, these data being given, How does the environment affect them, and how do they affect the environment? Now, I affirm that the relation of the visible environment to the great man is in the main exactly what it is to the "variation" in the Darwinian philosophy. It chiefly adopts or rejects, preserves or destroys, in short *selects* him. And whenever it adopts and preserves the great man, it becomes modified by his influence in an entirely original and peculiar way. He acts as a ferment, and changes its constitution, just as the advent of a new zoölogical species changes the faunal and floral equilibrium of the region in which it appears. . . . The mutations of societies, then, from generation to generation, are in the main due directly or indirectly to the acts or the examples of individuals whose genius was so adapted to the receptivities of the moment, or whose accidental position of authority was so critical that they became ferments, initiators of movements, setters of precedent or fashion, centers of corruption, or destroyers of other persons, whose gifts, had they had free play, would have led society in another direction.

(James 1880, 225–7)

The person that James had in his sights here was Herbert Spencer, with his laws of inevitable progress up to a definite goal – the kind of predetermination that is built into the notion that simplicity leads inexorably to complexity. For James, there has to be an element of chance, of "hereandnowness" as he might have said, in the change. This is the essence of Darwinism – and, we might also say, of pragmatism. Not only with individuals but with the very fabric of our minds and our thinking:

the new conceptions, emotions, and active tendencies which evolve are originally produced in the shape of random images, fancies, accidental out-births of spontaneous variation in the functional activity of the excessively instable human brain, which the outer environment simply confirms or refutes, adopts or rejects, preserves or destroys, – selects, in short, just as it selects morphological and social variations due to molecular accidents of an analogous sort.

(p. 247)

It is not so much that things are not better, but that there cannot be the absolute direction to perfection, including perfect knowledge, which was the aim of pre-Darwinian philosophy.

The evolution of theories

So far, you might think, a promising beginning to the use of evolution in epistemology. Alas, only a beginning. As noted in the Preface, the first part of the twentieth century was not a good time for evolutionary approaches to epistemology. Much of this was the fault of evolutionists. It was their business if they wanted to turn their science into a secular religion, but the rest of the world could be forgiven if it did not want to follow. However, philosophers too were partly responsible for the divide. The development of modern logic and the achievements of physics led people to think that new methods could solve old problems and that the radical moves demanded by evolution were not needed. And especially as evolution was seen by philosophers as linked to pragmatism, it was denounced. The highly influential Bertrand Russell was a leader in condemning America's contribution to epistemology, not just as a false theory of knowledge, but also as a corrupt ideology leading to power. Even worse, it left the way open for religion: "In a chapter on pragmatism and religion he [James] reaps the harvest. 'We cannot reject any hypothesis if consequences useful to life result from it.' 'If the hypothesis of God works satisfactorily, on the whole, in the widest sense, it is true'" (Russell 1945, chap. 29).

By the 1960s, things were starting to change. Evolutionary biology had by now hauled itself up from its too-popular status to that of a firmly empirically and theoretically based professional science, one that deserved respect outside its field. A number of thinkers started to speculate about the ways in which an evolutionary understanding could aid the search for an adequate theory of knowledge. There were (and still are) two major approaches, both of which have grown from nineteenth-century thinking, although now the intent is almost always Darwinian rather than anything else. The first picks up on a version of the kind of thinking that we saw in William James – a version, although not exclusively a Jamesian version, because, since it is fairly obvious, it is to be found (generally floated more as an idea than as a developed hypothesis) in others around that time, including both Charles Darwin and Thomas Henry Huxley. It proposes that one thinks of ideas or concepts or theories as the units of selection: that is to say, one thinks of a struggle between ideas or concepts or theories and consequently the success of some and the failure of others.

In this mode, in the *Descent* Darwin wrote with approbation of the German philologist and orientalist Max Müller on the origin of language:

As Max Müller has well remarked:– "A struggle for life is constantly going on amongst the words and grammatical forms in each language. The better, the shorter, the easier forms are constantly gaining the upper hand, and they owe their success to their own inherent virtue." To these more important causes of the survival of certain words, mere novelty and fashion may be added; for there is in the mind of man a strong love for slight changes in all things. The survival or preservation of certain favoured words in the struggle for existence is natural selection.

(Darwin 1871, 1: 60)

Darwin did not follow through on this, and in my judgment he was more committed to a different approach to evolutionary epistemology (as this is generally known), an approach we shall encounter in a moment. In recent times, one who tried to cast his theory of knowledge in Darwinian terms of this ilk was Karl Popper (1972, 1974). He argued that his theory of falsifiability, where the mark of the genuine scientific theory is that it be open to checking and (if the facts tell against it) falsification, is essentially Darwinian. You start with a problem, you offer a tentative solution to this problem – a bold conjecture – you open it up to checking and, if need be, rigorous refutation, and then you find yourself with this solution, or more likely a modified problem, on your hands.

$$P_1 \rightarrow TS \rightarrow RR \rightarrow P_2$$

It is easy to see how this could get even more Darwinian if you offer two tentative solutions, TS1 and TS2, to the same problem and then let them fight it out – catastrophism and uniformitarianism, for instance, to explain the Alps.

The English-born philosopher Stephen Toulmin developed this Darwinian picture in some great detail. He writes:

Science develops . . . as the outcome of a double process: at each stage a pool of competing intellectual variants is in circulation, and in each generation a selection process is going on, by which certain of these variants are accepted and incorporated into the science concerned, to be passed on to the next generation of workers as integral elements of the tradition.

Looked at in these terms, a particular scientific discipline – say, atomic physics – needs to be thought of, not as the contents of a textbook bearing any specific date, but rather as a developing subject having a continuing identity through time, and characterized as much by its process of growth as by the content of any one historical cross-section . . . Moving from one historical cross-section to the next, the actual ideas transmitted display neither a complete breach at any point – the idea of absolute 'scientific revolutions' involves an over-simplification – nor perfect replication, either. The change from one cross-section to the next is an evolutionary one in this sense too: that later intellectual cross-sections of a tradition reproduce the content of their immediate predecessors, as modified by those particular intellectual novelties which were selected out in the meanwhile – in the light of the professional standards of the science of the time.

(Toulmin 1967, 465–6)

A number of historians of science have embraced this philosophy of change. Notably, the American thinker David Hull wrote a history of recent turmoils in systematics, the theory and practice of the classification of organisms. The traditional approach to systematics, endorsed by the Grand Old Men of twentieth-century Darwinism, particularly the paleontologist G. G. Simpson and the ornithologist Ernst Mayr, was a somewhat ecumenical, evolutionary gallimaufry that took into account history, morphology, and anything else that seemed relevant. It was known, for example, that the birds and the crocodiles are fairly close relatives considered historically, but this approach separated them because they have gone off on very different evolutionary paths. Then first came new methods ("numerical taxonomy") using computers, which involved gathering huge amounts of information willy-nilly and grinding it up and spitting out results no matter what the evolution tells us. And second came the already discussed, German-founded approach called cladism (or "phylogenetic taxonomy"), which eschewed everything but past history and proposed refined techniques (also increasingly using the power of computers) to ferret out the paths of the past.

By digging into the archives, by looking at referees' reports, by seeing what actually got published and where, and above all by extensive interviewing of the main players, Hull argued that the progress of the dispute – from which the cladists emerged as the ultimate and total winners – lent itself entirely to an evolutionary analysis of the kind we are considering here. The various groups struggled and fought for their ideas, and slowly but surely the cladists won through to the top. New entrants to the field identified with the cladists and internalized and practiced that approach, or else. Exploiting the evolutionary analogy to the full, Hull argued that in biology we have selfish genes, working to maximize their own advantage. Likewise in science we have selfish scientists, working to maximize their own advantage. It is a kind of invisible-hand situation, where good science emerges from the self-interest of individuals. "Science is so organized that once a person who is curious about nature gains entry into a particular scientific community and begins to receive credit for his or her contributions, the system of mutual use and checking comes into play. Science is so organized that self-interest promotes the greater good" (Hull 1988, 357).

Scientists need each other for good ideas and also for help of a more immediate nature. On the one hand, there is the professor or the senior scientist. He or she has a good idea, but an idea is just an idea unless it is taken up. "Unless a scientist is content to play the role of an unappreciated precursor, he or she had best do more than just publish. Where to publish, in what form, how often, who to cite, who not to cite, and so on all enter into the process of making one's work public" (Hull 1988, 366). The senior scientist must have students or collaborators or others who are going to listen and be prepared to quote the work. However, no one is going to do this without good reason: the senior scientist provides grants or, even more significantly, furthers the careers of the others through citing their work or being prepared to add their names to publications and so forth. It is all a question of give and take. "Scientists treat exactly the same thesis differently depending on who presents it, whether friend or foe. A view presented by a friend is given the benefit of the doubt, and, if rejected, rejected gently and usually in private. The same view presented by someone outside one's circle of allies is greeted quite differently" (p. 390).

Critique

To the pragmatist, to the evolutionary epistemologist of this type, the proof of the pudding is in the eating. If this philosophy leads to interesting and insightful history, then this in itself is justification enough. There is no higher court of appeal. For the rest of us, this is all a little self-reinforcing, so we might legitimately ask about such a use of evolutionary ideas. Unless you share the metaphysics of a Spencer and perhaps a Peirce, in which everything is part of a cosmic whole moving towards the ultimate good or perfection, then the main point to note is that you are dealing with an analogy – scientific ideas or theories are not organisms – and it is proper to ask about the points of difference. The most obvious (assuming that one is working in a Darwinian framework) is that the variations of biology are random – not in the sense of being uncaused but in the sense of not appearing to be directed towards need – whereas the variations of knowledge (let us stay with science as our paradigm) are not. Charles Darwin knew full well that the origin of organisms was the "mystery of mysteries," and he labored hard to come up with the explanation. There was nothing random about his arrival at natural selection.

Added to this is the fact that the ideas can be transmitted sideways, as it were. In biology, if I develop a good variation, then you are out of luck, unless you can mate with me or one of my offspring which has this variation. And even then, it is not you but your offspring that might get the variation. In science, a good idea gets passed around the group and all can benefit at once. An idea might even transfer from one area of science to another. Paleontologists are very interested today in evolutionary development, wondering if it might hold the key to some of the changes they see in the fossil record. Third, hybridization may take place. This occurs often in the plant world, and many species today are the results of separated groups hybridizing. It happens less often in the animal world, and generally – as with the mule – points to the dead end of sterility. But whether it occurs in plants or animals, there is a general feeling that this is not the way of massive innovation, that it will not produce the really big breakthroughs that sometimes happen in evolution. In science, however, the coming together of ideas under one hypothesis – the consilience of inductions – is often, if not always, the mark of the really important. Plate tectonics, the movement of continents on their own slabs or plates, is a winner because so much is explained by the one basic idea. The same is true, as we know very well, of Darwinian evolutionary theory.

All of this suggests that, although it is surely true to think of the course of science as developmental – evolutionary if you will – it is a development very different from that to be found in biology. In a sense, science is directional, progressive, moving towards a goal that many would say is to provide a true description or reflection of objective reality. Darwinism works because evolution through natural selection really did occur. The same is true of plate tectonics. Phlogiston theory was rejected because phlogiston does not exist. Lamarckism was rejected because acquired characteristics are not inherited. Inasmuch as the work of Hull is insightful, it is not Darwinian, and inasmuch as it is Darwinian, it is not insightful.

Responses

There are three responses one can make here. First, that favored by Hull himself: one admits the disanalogies, one points out that analogies always have disanalogies, and one concludes that, if one gets some good results – and Hull did – that is enough for one day. So what if the two kinds of change are different? One does not have to claim that they are the same in order to exploit the analogy and to get some good results. Take the question of progress. Perhaps there are differences between biology and concepts, but there are enough similarities to keep the analogy going.

> Whenever the conditions are right, evolution by means of natural selection will occur. The global goal of natural selection may well be increased adaptation, but for particular lineages the particular goals keep changing, not because genetic variation is "blind," not because natural selection is non-intentional, but because so many aspects of the environment to which organisms must adapt keep changing. Conceptual evolution, especially in science, is both locally and globally progressive, not because scientists are conscious agents, not because they are striving to reach both local and global goals, but because these goals exist. Eternal and immutable regularities exist out there in nature. If scientists did not strive to formulate laws of nature, they would discover them only by happy accident, but if these eternal, immutable regularities did not exist, any belief that a scientist might have that he or she had discovered one would be illusory.
>
> (Hull 1988, 476)

A second defence of this kind of evolutionary approach to knowledge argues that, truly, Darwinism – at least, a revised and better

Darwinism – does approximate more closely to the conceptual-change model than has been suggested in this section. This was very much the position of Popper, who (at least for a good while) argued for a kind of theory of directed mutation, where changes occurred as needed and in the right direction. However, as you might imagine, even those Darwinians who favor some kind of evolutionary progress find this solution quite unacceptable. It has the odor of intention, of design, that is intolerable in modern evolutionary theory – indeed, in modern science. Popper was certainly not at all inclined to admit theological infiltration of science, but there is an uncomfortable resemblance between his view of the evolutionary process and that of those who welcome such God-based, intruding factors.

A third suggested approach goes rather the other way. In one sense, it would be hard to imagine a more non-evolutionary position than that view of science proposed by Thomas Kuhn in his celebrated work *The Structure of Scientific Revolutions* (1962). Kuhn argued that scientific work is embedded in what he called "paradigms," these being ways of viewing the world that provide scientists with challenges to solve (what he called "puzzles") and which, in the normal course of activity, go unchallenged. Every now and then a paradigm breaks down and (if a science is lucky) someone proposes a new paradigm within which work can recommence. The key thesis of Kuhn's book is that the switch from one paradigm to another is not fueled by pure reason but is more akin to a political or religious shift – now you see things one way, now you see things another way. In Kuhn's language, the two paradigms are incommensurable and there is a genuine break between the one and the other. It is not just that there are different ways of seeing the same world. The world itself has changed.

As I say, it is hard to imagine anything less evolutionary than this. There is change, but continuity is denied. Nevertheless Kuhn argued that in an important sense his theory of theories is Darwinian – namely, with respect to advance or progress. Kuhn argued that in biology we clearly get advance, but that it is not a simple matter of things getting better in an absolute way, but a matter of their getting better than what has gone before in a relative way. His view of paradigms was just this. There is no absolute progress – apart from anything else, Kuhn denied an objective world toward which science is advancing by ever closer approximations – but there is a kind of relative progress. You cannot go back, and the new paradigm in some way is able to function, whereas the old paradigm cannot.

> It helps to recognize that the conceptual transposition here recommended is very close to the one that the West undertook just a century ago. It is particularly helpful because in both cases the main obstacle to transposition is the same . . . The "idea" of man and of the contemporary flora and fauna was thought to have been present from the first creation of life, perhaps in the mind of God. That idea or plan had provided the direction and the guiding force to the entire evolutionary process. Each new stage of evolutionary development was a more perfect realization of a plan that had been present from the start.
>
> For many men the abolition of that teleological kind of evolution was the most significant and least palatable of Darwin's suggestions. The *Origin of Species* recognized no goal set by God or nature. Instead, natural selection, operating in the given environment and operating with the actual organisms presently at hand, was responsible for the gradual but steady emergence of more elaborate, further articulated, and vastly more specialized organisms. Even such marvelously adapted organs as the eye and hand of man – organs whose design had previously provided powerful arguments for the existence of a supreme artificer and advance plan – were products of a process that moved steadily from primitive beginnings, but toward no goal.
>
> (Kuhn 1970, 172)

As we have seen, the whole question of biological progress – especially biological progress in a post-*Origin* world – is complex and controversial. However, there are certainly many who would feel at least some approval of Kuhn's reading of the notion. Many arms race enthusiasts would fit in here. Their thinking generally focuses on a relative rather than absolute kind of progress. Advance has to be a kind of pragmatic success, a doing better than one's competitors. In Kuhn's case, there could be no such absolute in the Popperian sense, because for him there is no objective world against which to judge success. In this context, it is worth noting that some of the most successful history of science based on an evolutionary model is precisely that focusing on issues or approaches that are less about the world as such and more about interpretation. Although Hull seems to be a realist about the external world, his history of systematics is not about which organisms did or did not exist, but much more about the ways in which we might put them into groups. There is bound to be an element of subjectivity to an activity such as this. One set wants to record one kind of information and another set wants to record another kind of information. The success of the winners was less a matter of their having the only true method than of their having power and influence and so forth.

Having said this, let us not forget that, judged as an adequate account of science, Kuhn's theory has been much criticized in the years since it first appeared. Even if one can write a successful history of systematics using some theory that shares elements of Kuhn's thinking, one might question whether one could write a successful history of (say) physics using such a theory. As it happens, in the final chapter of this book I shall have more to say about Kuhn's approach, and in some respects what I shall say will be quite sympathetic. Nevertheless, it remains the case that in essence Kuhn's thinking is deeply anti-evolutionary – at least in any biological sense. It really is not genuinely Darwinian.

Innate dispositions

There is another way in which one can apply Darwinian thinking to problems of knowledge, a more literal approach. Herbert Spencer is a forerunner here. In his *Principles of Psychology* (1855), Spencer offered a kind of evolutionary Kantianism to explain how we know. We get information from without, but we structure it ourselves. This mental process of structuring is not, as Kant would argue, part of the necessary conditions for any kind of rational thought, but rather the result of the evolutionary process. Those proto-humans who structured reality in a certain way were better off than those that did not, and then habit and ingraining did the rest. Although many think (with Spencer) that this approach entails a kind of Darwinized Kantianism, in some respects (as I shall show) I prefer to think of it as the philosophy of David Hume brought up to date by the theory of the *Origin of Species*. Simply, one argues that there are certainly channels along which thinking must proceed and these structure and inform our knowledge. But rather than being just unexplained psychological tendencies, as they are in Hume, or necessary conditions for rational thought, as they are in Kant, they are dispositions put in place by our biology because they proved useful to our ancestors in the past. In other words, they are adaptations.

In one of his early notebooks, Darwin tried out this idea. "Plato . . . says in Phaedo that our 'imaginary ideas' arise from preexistence of the soul, are not derivable from experience.– read monkeys for preexistence" (M 128; Darwin 1987, 551). He does not seem to have followed it through in any detail after that, but, given what we shall see of his approach to ethics, my sense is that this approach is truer to Darwin's spirit than the struggle-between-ideas approach we have been considering above. One person in the last century who did take seriously the need to understand human cognition in terms of its evolutionary history was the ethologist Konrad Lorenz (1941), who argued that Darwin gives a biological backing to the philosophy of Kant. The mind is structured in certain ways to think and to act: this structure is a function of past evolutionary success, and hence we today are able to make sense of experience in ways that aid us to survive and reproduce. Lorenz – who was writing in German at the beginning of the Second World War and whose ideas therefore did not enter into the English-speaking world until many years later – seems to have shared Kant's views that the way we think is the only way in which we could think. We think causally because of our evolutionary past: to take an example, the person who sees a child crying because it is burnt by the fire, says that the fire caused the pain, because that is the thinking that led to survival. In some sense, we have to think causally because that is the only way that will lead to success – or, at least, that is the way that will lead to greatest success. Lorenz combined this with a belief in an objectively existing real world – the Kantian thing-in-itself – which knowledge is trying to map, although he also thought (in a very un-Kantian way) that his evolutionary approach allowed us to know about this reality.

By the 1960s, the pragmatic–naturalistic way of thinking in American philosophy was starting to reassert itself, and we find hints and suggestions, perhaps connected with the rise to respectability of evolutionary studies, that this kind of approach to knowledge may have virtues. One who threw out favorable hints about this idea was W. V. O. Quine, who wrestled with the fact that causal connection seems to have no justification and yet is the very thing on which we rely. When we argue "inductively," supposing that what happened in the past is a guide to what will happen in the future, we rely on assumptions about the regularity of the world.

> One part of the problem of induction, that part that asks why there should be regularities in nature at all, can, I think, be dismissed. That there are or have been regularities, for whatever reason, is an established fact of science; and we cannot ask better than that. Why there have been regularities is an obscure question, for it is hard to see what would count as an answer. What does make clear sense is this other part of the problem of induction: why does our innate subjective spacing of qualities accord so well with the functionally relevant groupings in nature as to make our inductions come out right? Why should our subjective spacing of qualities have a special purchase on nature and a lien on the future?
>
> There is some encouragement in Darwin. If people's innate spacing of qualities is a gene-linked trait, then the spacing that has made for the most successful inductions will have tended to predominate through natural selection. Creatures inveterately wrong in their inductions have a pathetic but praise-worthy tendency to die before reproducing their kind.
>
> (Quine 1969, 126)

Building on suggestions like these, and encouraged by the developments in evolutionary biology – especially the rise of human sociobiology – a number of people, including myself, tried to go all the way in putting together a theory of knowledge that started with the evolutionary legacy informing the way we think and act (Ruse 1986). In my own case, using a term introduced by Edward O. Wilson (writing with a young physicist, Charles Lumsden (1981)), I suggested that thinking follows biologically backed "epigenetic rules", although today I would perhaps prefer to use the simpler and more informative "innate dispositions" or "innate capacities." These are not the same as innate knowledge in the sense that John Locke demolished at the beginning of his *An Essay Concerning Human Understanding*, but the innate channels structuring our thinking that Locke does allow.

> For if we will reflect on our own ways of thinking, we will find, that sometimes the mind perceives the agreement or disagreement of two ideas immediately by themselves, without the intervention of any other: and this I think we may call intuitive knowledge. For in this the mind is at no pains of proving or examining, but perceives the truth as the eye doth light, only

by being directed towards it. Thus the mind perceives that white is not black, that a circle is not a triangle, that three are more than two and equal to one and two.

(Locke 1959, IV: ii, 1)

Augmenting Locke, we have at a minimum not only the basic rules of arithmetic and logic, but these are supplemented by what philosophers call "epistemic norms" or values, like consistency, simplicity, that which leads to predictive ability, and so forth (McMullin 1983). Consilience or unificatory power would rate highly here. There is no absolute reason why unificatory power should be a rule of human thinking, but only those of our would-be ancestors who took it seriously tended to become real ancestors. The proto-human who saw blood stains and footprints and heard growling in the undergrowth and said "tigers" and fled tended to live another day. The proto-human who saw blood stains and footprints and heard growling in the undergrowth and said "just a theory not a fact" and stayed tended not to live another day.

Although it is surely true that people with knowledge of science and use of technology tend – at least in the short run, which is all that counts for natural selection – to be more successful biologically than people without such science and technology, no one is making silly (and obviously false) claims that our preferring one particular scientific theory over another always is related to biological success. Mendel had no children whereas Darwin had seven that survived to maturity, but Mendel's thinking on heredity was more fruitful than Darwin's. The claims, rather, are that the science and technology (or other forms of knowledge) are based on, or informed by, rules of reasoning that have their roots in the biological struggle for existence and reproduction. In my case for an evolutionary epistemology, certainly, there is no claim that the rules we have are necessarily the only rules that one could have, although perhaps some such argument could be made for logic and parts of mathematics. But something like simplicity clearly is subjective in a sense, and rooted in evolutionary convenience – "Keep it simple stupid, or you die!" – rather than in eternal Platonic verities.

The same is true of much else. I can even imagine odd worlds where one might not have causation as we have it – perhaps fire is considered sacred or some such thing, rather than hurtful. Certainly, work by evolutionary psychologists on reasoning leads one to think that the exact rules of reasoning that we use are empirically based, reflecting the kinds of situations in which humans find themselves. Inferences that require exactly the same logic might be very easy for people when presented in familiar circumstances and very difficult for people when presented in unfamiliar circumstances.

> The coevolutionary dependency of truth standards on value applies to every component of our evolved neurocomputational architecture. The design of every system should have been impacted by this relationship. Because knowledge acquisition systems evolved to form the basis of action, the kinds of actions the system has evolved to engage in will build in different procedures for establishing truth criteria for different kinds of functions.
>
> (Tooby, Cosmides, and Barrett 2005, 224)

This is all very Humean in spirit rather than Kantian.

Culture?

The innate dispositions lead us to look at and conceptualize the world of experience in one set of ways rather than another. What about culture? It is certainly not a question of biology or culture but not both. Rather it is a matter of culture as produced, or informed, by biology. As you might expect from what has gone before, I argue strongly that metaphors play a major role in producing science, and these work with the dispositions to create science (Ruse 1999). Thus for instance, we want (biologically it paid us to want) theories that are going to be fruitful and lead to new kinds of predictions. The way in which we get these is often if not always through metaphor. Darwin and his predecessors and successors have used the metaphor of a division of labor to produce the kind of science that they desired. One might argue that in some respects the division of labor is part of our biology (though I am not sure that it is, given the success of Japanese auto workers in producing the goods by all sharing and cooperating in the tasks), but for Darwin and his successors this metaphor was part of their culture. This is how Darwin got it.

Darwinian epistemology is going to be a fusion of biology and culture. Does this mean that knowledge is relative? Well, certainly not necessarily across a particular society, perhaps not even across the human species. An alien race elsewhere in the universe might think differently from us – it might, if the certainty of the Kantian is gone – although it will not think in such ways as to contradict us. Creationism will not be the norm on Andromeda: that is to think our way but incorrectly. However, even if we hold the biology constant, there is certainly going to be some degree of cultural relativity. Suppose we had never had Christianity, with its interest in origins. It might be that today we would simply not think in evolutionary terms. We would not be Creationists, first because this demands thinking in post-Christian terms and second because it is wrong. But we might just not ask questions about origins – we might have the knowledge but cut it up other ways.

More precisely, if we had not had the metaphors of industrial Britain – struggle, division, even selection (not to mention design) – it is hard to imagine that we would ever have had a theory like Darwin's. This does not mean that Darwin's theory succeeded just because it was part of our culture. It succeeded because it triumphed when judged against the epistemic criteria of good science. Creationism – which is even more part of American culture than Darwinism – failed by these criteria. So yes, science is not absolute in the sense of one and only one possibility; and no, science is not relative in the sense that if it feels good then it must be right.

Many, probably most, evolutionary epistemologists think that one must be a realist, that science makes sense only if it is understood that there is a real world that one is mapping. I see nothing in what has been said that denies this possibility. There is never just one unique way to map or to picture something. Monet's picture of a cornfield is very different from Van Gogh's. But there is still a cornfield and a cornfield is the same for everyone. Someone who painted an auto graveyard while trying to picture the cornfield would be as mistaken as the Creationists. However, I am not convinced that one must be a realist to be an evolutionary epistemologist of the kind being discussed now. One could think that the very notion of an objective reality – the tree that falls in the forest when no one is around – is an absurdity, and the only sense we can make of reality is as something that is observed. With Kuhn, we argue that the reality and the perception of the reality are as one.

I am not now saying that either realism or non-realism is right or wrong. I am saying that neither realism nor non-realism is uniquely the position one must adopt if one accepts this theory of knowledge. I doubt that the kind of empirical discussion we have been having – about the ways in which biology and culture infuse knowledge – can ever decide between realism and non-realism. Any decision must be made on metaphysical or theological or some such grounds. Or remain unanswerable, as it is for me. The wonder is that we can know as much as we do. Given that we are apes who acquired adaptations to get out of the jungle and onto the plains, there is no good reason to think that every problem is soluble to organisms like us. Richard Dawkins makes this point: "Modern physics teaches us that there is more to truth than meets the eye; or than meets the all too limited human mind, evolved as it was to cope with medium-sized objects moving at medium speeds through medium distances in Africa" (Dawkins 2003, 19). Let us leave it at that.

Objections

What of objections? Some simply want to dismiss the whole idea that the mind is shaped by natural selection, at least in any way that reflects the needs of organisms in their thinking and acting. Stephen Jay Gould (2002) floated this idea, arguing that perhaps the whole human thinking apparatus is a spandrel, something that evolved but perhaps as a byproduct of other needs and selective pressures. To say the least, this seems highly unlikely. To stress again: brains are very expensive in biological terms. Big brains need lots of protein: that is, one needs to be a carnivore and that means effort and time. One has got to get meat. Getting meat takes intelligence and brains, in a kind of feedback process. Unless selectionists are completely off the track, the spandrel suggestion is no more than that.

Other critics take a different approach, arguing that the wonders and mysteries of knowledge simply could not have been revealed by naturally formed brains, however powerful selection might have been.

Perhaps selection could have got us to think that $2 + 2 = 4$, but it could never have led to brains capable of the discovery of the Euler equation:

$$e^{\Pi i} = -1$$

This is a popular argument with Christians (Polkinghorne 1991). Yet, as Richard Dawkins has said, repeatedly, it is always a dangerous argument to say that natural selection cannot have done something, and part of the problem here is to know what other hypothesis is being proposed. God is no solution, at least no scientific solution. Mathematicians, notoriously, are Platonists, believing that there is some world in which mathematical relationships hold eternally. If there is such a world, then, since brains can obviously access this world, why should brains produced through evolution have some special difficulty in accessing it? If, however, mathematics is something else, whatever that something else may be – marks on paper, a construction of some sort – again it is difficult to see why a brain produced by evolution should be barred from finding the difficult (and amazing) formulae. The problem seems more one for philosophy of mathematics than for philosophy of biology.

A more threatening attack on any kind of evolutionary approach to knowledge has been launched by the philosopher (and ardent Christian) Alvin Plantinga (1991). Like the spider with the fly, he welcomes us in, agreeing that if we are Darwinian evolutionists this must extend to our reasoning and cognitive powers. But then Plantinga shows his true arachnid nature and he is no Charlotte, I am afraid. He argues that Darwinian evolution cares nothing for truth, only for survival and reproductive success. As is sometimes said – not by Plantinga! – Darwinism is the science of the four effs. Fighting, fleeing, feeding, and reproduction. Thus, there is no reason why our reasoning and cognitive powers should tell us the truth about the world: all that they need do is tell us what we need to believe to survive and reproduce. And this information, so long as it is effective, could as easily be false as true. Plantinga tells the story of a posh dinner at an Oxford college, where Richard Dawkins spoke up for atheism before the philosopher A. J. Ayer. From this he draws a philosophical moral. Perhaps none of our thoughts can tell us about reality. Perhaps we are like beings in a dream world:

Knowledge

Their beliefs might be like a sort of decoration that isn't involved in the causal chain leading to action. Their waking beliefs might be no more causally efficacious, with respect to their behaviour, than our dream beliefs are with respect to ours. This could go by way of pleiotropy: genes that code for traits important to survival also code for consciousness and belief; but the latter don't figure into the etiology of action. It could be that one of these creatures believes that he is at that elegant, bibulous Oxford dinner, when in fact he is slogging his way through some primeval swamp, desperately fighting off hungry crocodiles.

(Plantinga 1993, 223–4)

Everything we believe about evolution could be false. And this is obviously to disprove Darwinian epistemology by a *reductio ad absurdum*. If our theory of knowledge can embrace indifferently the true and the false in the interests of expediency, we are in deep trouble. Plantinga calls this "Darwin's Doubt," because it was even expressed by a worried Darwin himself. "With me the horrid doubt always arises whether the convictions of man's mind, which have been developed from the mind of the lower animals, are of any value or are at all trustworthy. Would anyone trust in the convictions of a monkey's mind, if there are any convictions in such a mind?" (Plantinga 1993, 219, quoting Darwin 1887, 1: 316). As a matter of fact, Darwin immediately excused himself as a reliable authority on such philosophical questions, but this somewhat awkward point goes unanswered.

Is this a serious counter-argument? Darwinians agree that there are times when organisms and their characteristics will be out of adaptive focus. Genetic drift is a case in point, as is that phenomenon mentioned by Plantinga, "pleiotropy," where a single gene controls two different characteristics and a non-adaptive feature piggybacks on an adaptive feature. But biological possibility does not equal reasonable biological actuality. Plantinga has hardly described situations remotely like those that evolutionists think ever actually happen or are true. Almost by definition, drift has minor effects – so minor that selection never picks them up. Thinking that you are matching eminent philosophers drink for drink may be the way to conjure up pink rats, but it is not the way to fight off crocodiles. Fighting crocodiles requires force and defence and cunning and fear and split-second reaction to danger and much more. At the very most, the sorts of mechanisms mentioned by Plantinga are going to make you a bit uncertain about some things occasionally – some things that do not much matter anyway. If we need to know the truth, and facing crocodiles is a paradigm instance of such a need, drift will not stand in our way. It is not sufficiently powerful to make evolution through selection that maladaptive or ineffective.

Having said this, let us agree that – unreal example or not – Plantinga has seized on a question of some importance. It is true that, leaving non-adaptive causes on one side, there is sometimes deception for selection-related reasons. Sometimes, the deception is systematic, as we all learn in elementary psychology classes. However, such cases of deception are really not that mysterious or inexplicable. Either (as so often in the psychology class examples) the cases of deception are those we would never have encountered in our past, and hence there is no reason to expect Darwinian built-in safeguards against them, or Darwinian evolutionary theory itself can give good reasons why the deceptions occur. Go back to the objective necessity of causal connection. Hume showed that this is an illusion. Why do we have it? Because those proto-humans who associated fire with burning survived and reproduced, and those who thought it was all a matter of philosophy did not. Selection itself explains why we think that causes really do exist as entities.

The point about systematic misconceptions is that we can work out why they occur and how it is that they come from selection. There are cases where no misconceptions can occur – at least, none where it would be in selection's interests that we be properly informed. Falling trees hurt; drinking arsenic kills; beautiful naked people (with provisos about sex and orientation) are a sexual turn-on; roses really do have thorns that prick. These are not and cannot be cases of misconception. Selection is not about to kid us on such a topic as the hurtfulness of falling trees. The world is not crazy. Plantinga's example of fighting crocodiles while you think you are in Oxford talking wise thoughts is truly crazy. The deceptions of selection work, make sense, just as do the non-deceptions. We cannot be deceived all of the time. More than this: because evolution generally leads to true conclusions, we can use these to ferret out the counter-examples – those instances where selection does deceive us.

Plantinga will continue to press his argument. What of our touchstones? Is it not possible that we are deceived every moment of the day and night? Given the known delusions of love, who dare say that our genitalia are true guides to reality? Could we not be like the prisoners in Plato's cave in the *Republic*, thinking that shadows on the wall are the real thing? Plantinga invites us to think of a factory where everything seems red. To the person outside, it is all a question of filters. The ignorant and deceived workers have spectacles that color their perceptions. In truth nothing is red. But the factory worker has no genuine touchstone by which to make a judgment. No matter what care he takes in his assessments he will be mistaken. Could we humans not be all in the same situation? Could it not be that everything in the world/factory is false/red-seeming-but-not-really?

In response, the evolutionist can reply only that possibly we will never know the whole story and may be mistaken about any detail. However, there is a difference between the factory fiction and our experienced existence, suggesting that in real life we cannot be mistaken about all of the details. The factory example breaks down as an appropriate analogy precisely because, at some point, someone gets out to find that the redness is filter-caused. In real life, we can never check that everything is true-seeming-but-not-really. We can never get outside our evolution-produced bodies. So in a way it is difficult to know precisely what one might mean by saying that our thoughts are all totally mistaken. The case is identical to the thought experiment that has in recent years given so many philosophers so many happy hours of argument and speculation. Could we all truly be brains in a vat, wired up for sensations and kept alive by the right combination of chemicals and nutrients? It seems at first sight that this is at least a logical possibility, but in truth the more you think about it the less likely it seems even to make sense (Putnam 1981). We know what real brains and real vats are like, but these kinds of philosophical brains and vats may not make sense. Apart from anything else, why should they in turn not be functions of brains and vats one degree higher? And these in turn are brains and vats . . . ?

Conclusion

The pragmatist laughs at all of this. Undetectable brains and vats are as nothing to him or her. And the evolutionist feels the same way. Which does bring us back to the point at which we entered this chapter. Obviously there is a pragmatic element to the knowledge theory of the evolutionist. If the metaphysical realist – the person who is really committed to the world of objective reality – insists that the evolutionist allows that we could always be mistaken, the evolutionist finally (somewhat wearily) has to agree. In a way, real world or not, the evolutionist is a coherent theorist of truth – believing that what counts is that things hang together and work – rather than a correspondence theorist of truth – believing that what counts is getting isomorphic with something out there. In the end, what counts is getting along, day by day. But the Darwinian knew that already.

9
Morality

We turn now to the other side of philosophy: morality. There are two big questions that philosophers ask. What should I do? Why should I do what I should do? The first question frames "normative" or "substantive ethics," and the second (to do with justification) frames the central question of "metaethics." Thus, to give an example, Christians at the normative level accept the Love Commandment: "Love your neighbor as yourself." At the metaethical level, many Christians accept some form of the Divine Command theory: "You ought to do that which is God's will simply because it is God's will." Giving simple answers like these does not end ethical discussion. Who is my neighbor? Why should I do what God wants? But the division does help to structure the discussion, even though no one pretends that the two levels are entirely separate.

Darwin on morality

Charles Darwin discussed morality in the *Descent of Man*. He did not give an answer at the metaethical level. This is not to say that we cannot get an idea of what he thought a proper approach should be. But Darwin stuck to his last as a scientist, thinking that his job was to show how morality comes into being. He therefore had to make some commitment at the normative level – What is morality? – and (in line with most of his countrymen at the time) Darwin opted for a kind of utilitarianism: the Greatest Happiness for the Greatest Number. However, he did want to put things more in terms of biological needs.

"The term, general good, may be defined as the term by which the greatest possible number of individuals can be reared in full vigour and health, with all their faculties perfect, under the conditions to which they are exposed" (Darwin 1871, 1: 98).

Darwin did not make a great deal of notions that are dear to Americans, like rights, although his discussion and endorsement of capitalism showed that he agreed to the right to have property and so forth. My suspicion is that his attitude reflected national prejudice – remember that Jeremy Bentham referred to rights as "nonsense on stilts." Essentially, Darwin was thinking in terms of a kind of generally accepted, normative morality. He saw religion as having a role in enforcing morality, so one suspects that he would have happily endorsed many if not most of the dictates of the New Testament. Things like rights are not so much absolutes – if they are absolutes, they are going to be difficult for an empiricist to handle anyway – as a consequence of the commonsense, English-middleclass-gentleman morality that Darwin accepted. Slavery is wrong because it makes for unhappiness: a slave cannot develop as fully as a human being as someone who is free.

At the level of which he was talking, Darwin was fairly sophisticated in his analysis of morality. He recognized that clearly we humans have selfish or self-directed desires. But he saw – perhaps his reading of Kant (Darwin had read the *Metaphysics of Morals*) had helped him here – that, without something more, such beings as humans cannot form societies. We have to have some sense or feeling of sociability, some felt need to be on good working terms with our fellow humans. There is a bit of a chicken-and-egg situation here. Are we social because we need to get together, or can we get together because we have the ability and need to be social?

It has often been assumed that animals were in the first place rendered social, and that they feel as a consequence uncomfortable when separated from each other, and comfortable whilst together; but it is a more probable view that these sensations were first developed, in order that those animals which would profit by living in society, should be induced to live together, in the same manner as the sense of hunger and the pleasure of eating were, no doubt, first acquired in order to induce animals to eat. The feeling of pleasure from society is probably an extension of the parental or filial affections, since the social instinct seems to be developed by the young remaining for a long time with their parents; and this extension may be attributed

Morality

in part to habit, but chiefly to natural selection. With those animals which were benefited by living in close association, the individuals which took the greatest pleasure in society would best escape various dangers, whilst those that cared least for their comrades, and lived solitary, would perish in greater numbers. With respect to the origin of the parental and filial affections, which apparently lie at the base of the social instincts, we know not the steps by which they have been gained; but we may infer that it has been to a large extent through natural selection.

(Darwin 1871, 1: 80)

The moral sense (as we might call it) is a product of evolution, primarily evolution through natural selection. Darwin devoted much time to showing how this social sense is something possessed by the animals and so might have been expected to evolve. But Darwin did not think that humans are just animals. We are fully moral – uniquely moral – because we have the ability to think about our actions, to judge them, to try to influence ourselves with respect to future behavior. In short, we have a conscience, capable of what we today might want to call "second-order desires", which sort through the first-order desires. At the first order, I want to look to myself, and also as a social being I want to look to others. I look to myself. Now I think about it, and am disgusted with myself and want not to have this feeling again. So the next time I try to do better. "The following proposition seems to me in a high degree probable – namely, that any animal whatever, endowed with well-marked social instincts, would inevitably acquire a moral sense or conscience, as soon as its intellectual powers had become as well, or nearly as well developed, as in man" (1: 71–2).

This is all rather Kantian sounding, and Darwin backed up his thinking with a purple-prose passage from the sage of Königsberg. "Duty! Wondrous thought, that workest neither by fond insinuation, flattery, nor by any threat, but merely by holding up thy naked law in the soul, and so extorting for thyself always reverence, if not always obedience; before whom all appetites are dumb, however secretly they rebel; whence thy original?" (1: 70). But, really, Darwin was on a very different, British empiricist track. For Kant, morality as we know it is a necessary condition for rational beings living together. Darwin calmly suggested that, had evolution gone another way, we might be rational but with a very different kind of normative morality.

It may be well first to premise that I do not wish to maintain that any strictly social animal, if its intellectual faculties were to become as active and as highly developed as in man, would acquire exactly the same moral sense as ours. In the same manner as various animals have some sense of beauty, though they admire widely-different objects, so they might have a sense of right and wrong, though led by it to follow widely different lines of conduct. If, for instance, to take an extreme case, men were reared under precisely the same conditions as hive-bees, there can hardly be a doubt that our unmarried females would, like the worker-bees, think it a sacred duty to kill their brothers, and mothers would strive to kill their fertile daughters; and no one would think of interfering. . . . The one course ought to have been followed, and the other ought not; the one would have been right and the other wrong.

(1: 73–4)

As in the case of epistemology, I think that Darwin was thinking and writing more in the tradition of David Hume than of any continental thinker, which is, of course, what one might expect. Note that I write "in the tradition" rather than "under the direct influence." Charles Darwin's debt to David Hume is a topic much discussed. I believe that Darwin wrote in the shadow of Hume's genius – as he wrote in the shadows of many other members of the Scottish Enlightenment, like Adam Smith; but, although there were some points of direct influence (which I shall touch on later), and for all that I locate Darwin's moral thinking in the Humean tradition, generally one should not seek to tie the bonds too close. More direct influences were figures like William Whewell, who was anything but a British empiricist. Darwin certainly relied heavily on that key notion, in Hume and others, of "sympathy", the emotion that lets you put yourself in the position of others when making moral judgments. Darwin certainly knew that this was Hume's term, but jottings that he made in the private notebooks he kept when inquiring into origins at the end of the 1830s, suggest that the direct influence came secondhand through the survey by James Mackintosh in *Dissertation on the Progress of Ethical Philosophy*, edited and with a preface by none other than William Whewell. Moreover, in those notebooks, Darwin refers to the usage of the term "sympathy" by others, including Adam Smith and Edmund Burke.

Whatever the strength of the Hume–Darwin links, they were there. Darwin wrote: "The aid which we feel impelled to give to the helpless is mainly an incidental result of the instinct of sympathy, which was originally acquired as part of the social instincts, but subsequently rendered . . . more tender and more widely diffused" (Darwin 1871, 1: 168). Remember that Hume thought that sympathy could be found in the animals – a very Darwin-anticipatory sentiment, as one might say. The notebooks show that in the summer of 1839 Darwin read Hume's essay "Of the reason of animals." He notes: "'Tis evident, that sympathy, or the communication of passions, takes place among animals, no less than among men. Fear, anger, courage, and other affections are frequently communicated from one animal to another, without their knowledge of that cause, which produc'd the original passion." (N 101; Darwin 1987, 592). Both Hume and Darwin use the example of surgery to illustrate their thinking about sympathy. Hume: "Were I present at any of the more terrible operations of surgery, 'tis certain, that even before it begun, the preparation of the instruments, the laying of the bandages in order, the heating of the irons, with all the signs of anxiety and concern in the patient and assistants, wou'd have a great effect upon my mind, and excite the strongest sentiments of pity and terror" (Hume 1965, 576). Darwin: "Nor could we check our sympathy, even at the urging of hard reason, without deterioration in the noblest part of our nature. The surgeon may harden himself whilst performing an operation, for he knows that he is acting for the good of his patient" (Darwin 1871, 1: 169).

In this last-quoted comment by Darwin, it is hard to imagine a more forceful defense of Hume's famous dictum: "Reason is, and ought only to be, the slave of the passions, and can never pretend to any other office than to serve and obey them" (Hume 1965, 415). But let me stress again what was said above. However insightful the philosopher might find Darwin, he was not writing as a philosopher. He was writing as a scientist who is trying to understand the nature and origin of that human dimension that we call morality.

Social Darwinism

So now let us move on to those more directly concerned with what we call the philosophical issues. As in epistemology, there have been and still are two approaches, and, not surprisingly, there are considerable parallels between the theories of knowledge and the theories of morality. We will begin with the more traditional approach to an evolution-infused attitude to morality, known as "Social Darwinism," although, as it happens, it owes more in spirit and content to Herbert Spencer than it does to Darwin. Starting with the normative level, the moves seem no less obvious than in the realm of knowledge. Ferret out the evolutionary process. Transfer it to the human realm. Argue that it is morally obligated. And there you have your moral directives. Spencer is well known, some would say notorious, for his tough stand on social issues, and these – commonly known as "laissez-faire" – seem to be a straight transfer from biology and the Darwinian process of struggle and selection.

> We must call those spurious philanthropists, who, to prevent present misery, would entail greater misery upon future generations. All defenders of a Poor Law must, however, be classed among such. That rigorous necessity which, when allowed to act on them, becomes so sharp a spur to the lazy and so strong a bridle to the random, these pauper's friends would repeal, because of the wailing it here and there produces. Blind to the fact that under the natural order of things, society is constantly excreting its unhealthy, imbecile, slow, vacillating, faithless members, these unthinking, though well-meaning, men advocate an interference which not only stops the purifying process but even increases the vitiation – absolutely encourages the multiplication of the reckless and incompetent by offering them an unfailing provision, and *discourages* the multiplication of the competent and provident by heightening the prospective difficulty of maintaining a family.
>
> (Spencer 1851, 323–4)

Actually, however, things are all a bit more complex than this. In fact, Spencer wrote this passage before he was a declared evolutionist, and in respects it owes most to his nonconformist background and his hatred of established privilege. He is truly arguing for a society that removes the barriers to success or failure and therefore lets the meritorious rise up. A hundred years later, this is just the philosophy on which the future British prime minister Margaret Thatcher was raised. Later in life, far from being that tough-minded, Spencer was to argue strenuously against militarism, which he saw as a waste of resources and a barrier to free trade. More significantly, as we know, Spencer himself always had a bit of a love–hate relationship with pure

Darwinism: as the chief mechanism of change, he favored the inheritance of acquired characteristics over natural selection. But all of this said, Spencer certainly was deeply committed to a naturalistic evolutionary approach to ethics, and he certainly did sometimes come across as a brute.

> Besides an habitual neglect of the fact that the quality of a society is physically lowered by the artificial preservation of its feeblest members, there is an habitual neglect of the fact that the quality of a society is lowered morally and intellectually, by the artificial preservation of those who are least able to take care of themselves . . . For if the unworthy are helped to increase, by shielding them from that mortality which their unworthiness would naturally entail, the effect is to produce, generation after generation, a greater unworthiness.
>
> (Spencer 1873, 343–4)

Echoes, and more, of this can be found in the thinking of Spencer's many followers, including those in America. Certainly, in the New World, one could find people who were happy to read their brutal industrial practices as manifestations of the evolutionary process and hence in some sense natural. Yet, here too one needs to take care. One great enthusiast for Spencer's thought was the Scottish-American industrialist Andrew Carnegie, the founder of US Steel. He was a major philanthropist, giving a great deal of money to support the founding of public libraries. This was part and parcel of his version of Social Darwinism, which – for him, as for many rich and successful men – stressed the success and superior qualities of the winners rather than the lack of success and inferior qualities of the losers. A public library was a place where a poor but gifted child could go and, through self-motivated effort, gain learning and raise him- or herself up from the masses (Russett 1976).

More generally, an ethics backed by evolution became a crucial part of what I have referred to as the turn of evolutionary thinking from straight science towards its late-nineteenth-century status as a kind of secular religion. And as we find disputes and differences in more conventional religions – today's evangelicals are against homosexuality, whereas more liberal groups like the Unitarians and Quakers find it morally unproblematic – so we find that there were strong differences between evolutionary ethicists in the years after Darwin. On the social front, for instance, we have seen how Darwin himself was happy to endorse capitalism as having biological backing. Remember: "the inheritance of property by itself is very far from an evil; for without the accumulation of capital the arts could not progress; and it is chiefly through their power that the civilised races have extended, and are now everywhere extending their range, so as to take the place of the lower races" (Darwin 1871, 1: 169). Alfred Russel Wallace (1900), on the other hand, the co-discoverer of natural selection, was a life-long socialist and thought that socialism was justified by biology. And then around the turn of the century, there was the Russian anarchist Prince Peter Kropotkin (1902), who also justified his political philosophy in the name of evolution. Like most Russian evolutionists, he saw the real struggle for existence as occurring, less between organisms and organisms, and more between organisms and the harsh environment. Evolution has put in place sentiments promoting "mutual aid" to help organisms survive in the face of harsh natural conditions.

> Two aspects of animal life impressed me most during the journeys which I made in my youth in Eastern Siberia and Northern Manchuria. One of them was the extreme severity of the struggle for existence which most species of animals have to carry on against an inclement Nature; the enormous destruction of life which periodically results from natural agencies; and the consequent paucity of life over the vast territory which fell under my observation. And the other was, that even in those few spots where animal life teemed in abundance, I failed to find – although I was eagerly looking for it – that bitter struggle for the means of existence, among animals belonging to the same species, which was considered by most Darwinists (though not always by Darwin himself) as the dominant characteristic of struggle for life, and the main factor of evolution.
>
> (Kropotkin 1955, vi)

Instead, Kropotkin saw help given by one animal to another. On the one hand, there were the harsh conditions:

> On the other hand, wherever I saw animal life in abundance, as, for instance, on the lakes where scores of species and millions of individuals came together to rear their progeny; in the colonies of rodents; in the migrations of birds which took place at that time on a truly American scale along the Usuri; and especially in a migration of fallow-deer which I witnessed on the Amur, and during which scores of thousands of these intelligent animals came together from an immense territory, flying before

the coming deep snow, in order to cross the Amur where it is narrowest – in all these scenes of animal life which passed before my eyes, I saw Mutual Aid and Mutual Support carried on to an extent which made me suspect in it a feature of the greatest importance for the maintenance of life, the preservation of each species, and its further evolution.

(p. viii)

Other social issues

We find the same divisions over the issue of force and the military. German thinkers, particularly, were convinced that Darwinism equals war equals that which is morally good or acceptable. "Struggle is . . . a universal law of Nature, and the instinct of self-preservation which leads to struggle is acknowledged to be a natural condition of existence. 'Man is a fighter'" (Bernhardi 1912, 13). And: "might gives the right to occupy or to conquer. Might is at once the supreme right, and the dispute as to what is right is decided by the arbitration of war. War gives a biologically just decision, since its decisions rest on the very nature of things." Hence: "It may be that a growing people cannot win colonies from uncivilized races, and yet the State wishes to retain the surplus population which the mother-country can no longer feed. Then the only course left is to acquire the necessary territory by war" (p. 15). On the other hand, there were those who thought that war is biologically stupid. Until his experiences in the First World War made him change his thinking dramatically, Vernon Kellogg, professor at Stanford and popular writer on evolution, argued strenuously that evolution points away from conflict and fighting.

Man is an incident of organic Evolution, and at bottom and in all things his body and his nature are the product of this Genius of Life. And just as Evolution made him, with his need, a Fighter, and taught him War, so now, with the passing of this need, with the substitution of reason and altruism for instinct and egoism, Evolution will make him a Man of peace and goodwill, and will take War from him. And any man will find his greatest advantage and merit in aiding, rather than delaying this beneficence.

(Kellogg 1912, 140–1)

Parenthetically, there is still much discussion between scholars about the extent to which Social Darwinism fed into the National

Socialist movement (Gasman 1971; Kelley 1981). There is no doubt that Hitler picked up some gleanings from it, and there are certainly passages in *Mein Kampf* that sound very much like the sort of thinking we have seen above. On the other hand, there were pressing reasons why the Nazis were not about to embrace Darwinism wholesale, starting with the fact that it stresses the unity of humankind – Jews and Gentiles are one group – and going on to the fact that even the most beautiful of Arians have a simian ancestry.

Let me mention one other topic of great interest to the late Victorians – again a subject on which Darwinians divided: the nature and status of women. As we know, Charles Darwin was pretty conventional on this issue, expressing views that seemed to come straight out of a Dickens novel. Remember: "Man is more courageous, pugnacious, and energetic than woman, and has a more inventive genius" (Darwin 1871, 2: 316). Balancing this, woman has "greater tenderness and less selfishness" (p. 326). Apparently, these characteristics are largely the consequence of the bravest of the male savages getting first pick of the prettiest female virgins available as mates. Relying on the steamy reports of Victorian explorers like Richard Burton (who wrote material of a kind that a gentleman like Darwin thought prudent to put into Latin, lest the books fall into the hands of children or servants), we learn that the reason for the big bottoms of the Hottentots is that the women are lined up and the first choices are made of those who protrude farthest *a tergo*. "Nothing can be more hateful to a negro than the opposite form" (p. 346, quoting Burton).

Wallace on the other hand was an ardent feminist, thinking indeed that the salvation of the human race lies in the hands (and minds) of women, young women particularly. Apparently in the future, men will no longer have much say in who mates with whom. They will be chosen as partners on their virtues, and these apparently do not include many of the traits that predominate today.

> In such a reformed society the vicious man, the man of degraded taste or of feeble intellect, will have little chance of finding a wife, and his bad qualities will die out with himself. The most perfect and beautiful in body and mind will, on the other hand, be most sought and therefore be most likely to marry early, the less highly endowed later, and the least gifted in any way the latest of all, and this will be the case with both sexes.
>
> (Wallace 1900, 2: 507)

With writing like this, Wallace showed as little insight into the true nature of young people as have those who write the behavioral codes of today's American liberal arts colleges, but he at least had some claim to understanding human biology.

The Social Darwinism movement had a bad odor by the beginning of the twentieth century: too often it was linked to harsh and cruel moral prescriptions, and so generally people denied connection with it. But it did continue to flourish, under other names or no names at all. Eugenics, the attempt to improve humankind by selective breeding or to prevent degeneration by barring some breeding, throve in many parts of the world, especially America, and then, with truly vile consequences, in Germany. More enlightened attempts at improving the lot of humans on evolutionary principles perhaps got less notice, but they too existed. Julian Huxley, the grandson of Thomas Henry Huxley, was a great enthusiast. The 1930s saw a dreadful world recession, and this led him to promote the New Deal's massive public works, like the Tennessee Valley Authority's project to dam rivers in order to bring electricity to the South. Julian Huxley had to be circumspect, for this was the time when Hitler was likewise promoting large state-financed projects (like the building of the Autobahns). But Huxley left little doubt about his feelings.

> All claims that the State has an intrinsically higher value than the individual are false. They turn out, on closer scrutiny, to be rationalizations or myths aimed at securing greater power or privilege for a limited group which controls the machinery of the State.
>
> On the other hand the individual is meaningless in isolation, and the possibilities of development and self-realization open to him are conditioned and limited by the nature of the social organization. The individual thus has duties and responsibilities as well as rights and privileges, or if you prefer it, finds certain outlets and satisfactions (such as devotion to a cause, or participation in a joint enterprise) only in relation to the type of society in which he lives.
>
> (Huxley 1943, 138–9)

Today, the crises we face come almost from too successful a society, from the ways in which the needs of the West, the USA particularly, lead to our raping the rest of the world to supply them. One thinks particularly of such things as the destruction of the Brazilian rain forests for short-term gain. Edward O. Wilson, entomologist and sociobiologist, has been a leader in crying out against this terrible destruction. But he does not argue just from intuition. He believes that humans have evolved in symbiotic relationship with the rest of nature, and that in a world of plastic, quite literally, we would die. This is almost an aesthetic response, although Wilson does argue also that we need biodiversity, as is provided by the rain forests. Who knows what medicines and similar vital products they might yield in the future? In a recent book, *The Future of Life*, Wilson declares: "a sense of genetic unity, kinship, and deep history are among the values that bond us to the living environment. They are survival mechanisms for us and our species. To conserve biological diversity is an investment in immortality" (Wilson 2002, 133).

Justification

Let us turn now to the other side of the coin. Why should anyone take seriously an evolution-based ethics? I mean, what justification is being offered? You might well agree that capitalism or socialism is a good thing, but why accept either, even if one or the other does have a connection with the mechanisms of evolutionary change? Why be for or against war because of evolution? Or massive state works or biodiversity? I suspect most people today do deplore the loss of the rain forests, but what does evolution have to do with justifying pre-scriptions about proper action with respect to the rain forests? The answer of philosophers to this question – the metaethical question of the way in which evolution justifies normative ethics – has been resoundingly negative. Following David Hume, they point out that there is a basic difference between claims of fact and claims of obli-gation. And without some supporting link – which is virtually never supplied – you cannot and should not go from one to the other. "Evolution does such and such" and "you ought to do such and such" are simply not the same claims. The early-twentieth-century philoso-pher G. E. Moore (1903) referred to making this kind of connection as committing the "naturalistic fallacy." You cannot derive moral statements, statements about values, from factual statements, statements about the way that the world is.

My experience is that enthusiasts for evolutionary ethics – Edward O. Wilson (with whom I once wrote a paper on morality and its connections to biology) is a paradigm case – are supremely unworried by the criticisms of the philosophers. Nothing new, you might say. Few are worried by the criticisms of the philosophers. In this case, however, they do have half a point. Why should it be wrong to derive "ought" from "is"? Perhaps it is wrong, but you need some argument. Wilson particularly points out that often in science we go from talk of one kind to talk of another kind. We start by talking about molecules and end by talking about gas pressure and temperature. This is acceptable. So why not evolutionary ethics? Well, you can respond, it is acceptable in gas theory, to take this example, but not in ethics because the inference is clearly illegitimate. I might have a strong desire to have sex with my neighbor's beautiful wife, but it does not follow that such an action would be morally acceptable. Agreed, says Wilson, but there is one exception to all of this – evolution! Generally, you cannot go from the way things are to the way things ought to be, but in the case of evolution you can. It is permissible to go from "this is the way that evolution has made things or drives things" to "this is the way that it should be" (Wilson 1984, 2006).

One starts to suspect that there is a hidden premise or rule of inference here – something akin to identifying temperature with the speed of molecules – and as one looks at the writings of evolutionary ethicists, past and present, one soon realizes that there is. It is our old friend progress. Evolutionary ethicists think that the course of evolution itself confers value: one goes from the molecule to the man, and, as one does this, one gets ever greater value. Hence, it is morally acceptable – morally obligatory – to work to preserve the evolutionary process if not to help it outright. It is not that the enthusiast for laissez-faire has anything against widows and children, but that without harsh measures everything will collapse and all will suffer. (One certainly heard strong echoes of that theme in the speeches of Margaret Thatcher.) Likewise, Julian Huxley and Edward O. Wilson have both been very strong supporters of the idea of evolutionary progress and, for both of them, their moral prescriptions are part and parcel of this support. Without large-scale public works, without biodiversity, the human race will wither and perish. This in itself is justification for action.

We have been through the whole progress business in detail already, so there is no need to start again here. Even if you can find some satisfactory measure of progress, I doubt that this can be a value-impregnated measure – certainly not a measure that gives absolute values. I put humans at the top, but it is I who am doing it, not evolution through natural selection. So let us conclude here. Traditional evolutionary ethics, Social Darwinism, has a bad name. This is partly deserved. But there is more to be said than simple condemnation. At the normative level, much that has been proposed has been admirable: like Christianity, some good and some bad, but by no means all bad. At the metaethical level, what has been proposed is less successful than its supporters claim, but (I do not intend to damn with faint praise) far from obviously stupid and, indeed, a guide to a better approach. If our moral nature is the end point of our evolution, this counts for something. Even if we cannot get traditional justification, it cannot all be pointless. Our evolved nature must count for something. Hume thought it did, even though he was the one who denied justification. So let us now see how we might try a different approach.

Darwin and the evolution of morality

Let us start with the science. Prima facie, Darwinism does not seem likely to be very helpful in studying morality. You might think that it is only after we have escaped from biology into culture that morality can start to kick in. This seems to be the position of Thomas Henry Huxley in a passage I quoted a couple of chapters ago. Of course, biologically we have to be animals that give rise to and can support culture, but the world of Darwin seems to be at best one step removed. Yet, as we have also seen, there are those – Wallace and Kropotkin, for instance – who thought that evolution and morality are much more intimately connected. The problem is (and this also takes us back to a discussion of an earlier chapter) Wallace and Kropotkin seem to have a false view of the evolutionary process. In a word, they are ardent group selectionists, believing that evolution can cherish and form adaptations for the group in face of the needs of the individual. And these days, most people do not want to go that way. Or let me rephrase this. Most people do not want to go that way unless there are compelling reasons. So the question really is whether we can show that individual selection can promote moral-like abilities in animals, humans specifically, or if for some reason group selection (which no one wants to say is contradictory) can operate and produce morality.

Charles Darwin himself was particularly torn on this issue (Ruse 1980). We know that he was deeply committed to an individualist perspective. By the time it came to the *Descent*, and the evolution of morality, Darwin was really in a jam. He accepted fully that morality is a group characteristic. That is the whole point about morality. He accepted also that morality seems to go against individual interest.

> It is extremely doubtful whether the offspring of the more sympathetic and benevolent parents, or of those who were the most faithful to their comrades, would be reared in greater numbers than the children of selfish and treacherous parents of the same tribe. He who was ready to sacrifice his life, as many a savage has been, rather than betray his comrades, would often leave no offspring to inherit his noble nature. The bravest men, who were always willing to come to the front in war, and who freely risked their lives for others, would on an average perish in larger numbers than other men. Therefore it seems scarcely possible (bearing in mind that we are not here speaking of one tribe being victorious over another) that the number of men gifted with such virtues, or that the standard of their excellence, could be increased through natural selection, that is, by the survival of the fittest.
>
> (Darwin 1871, 1: 163)

So what is the solution? In part, Darwin anticipated what has been called "reciprocal altruism." You scratch my back and I will scratch yours.

> In the first place, as the reasoning powers and foresight of the members became improved, each man would soon learn that if he aided his fellow-men, he would commonly receive aid in return. From this low motive he might acquire the habit of aiding his fellows; and the habit of performing benevolent actions certainly strengthens the feeling of sympathy which gives the first impulse to benevolent actions. Habits, moreover, followed during many generations probably tend to be inherited.
>
> (pp. 163–4)

But then Darwin did bite the bullet and get into something that seems to be promoted by group selection. People are moved by the praise and condemnation of their fellows.

> But there is another and much more powerful stimulus to the development of the social virtues, namely, the praise and the blame of our fellow-men.

The love of approbation and the dread of infamy, as well as the bestowal of praise or blame, are primarily due, as we have seen in the third chapter, to the instinct of sympathy; and this instinct no doubt was originally acquired, like all the other social instincts, through natural selection.

(p. 164)

It is hard to see how this could be promoted by other than a group-favoring process. If I do not worry about praise and blame, then I might act selfishly and have more offspring. It seems that it is only if the worry, which is for the good of the group, can be promoted by selection for the group that it gets preserved.

To do good unto others – to do unto others as ye would they should do unto you, – is the foundation-stone of morality. It is, therefore, hardly possible to exaggerate the importance during rude times of the love of praise and the dread of blame. A man who was not impelled by any deep, instinctive feeling, to sacrifice his life for the good of others, yet was roused to such actions by a sense of glory, would by his example excite the same wish for glory in other men, and would strengthen by exercise the noble feeling of admiration. He might thus do far more good to his tribe than by begetting offspring with a tendency to inherit his own high character.

(p. 165)

Having said this, however, Darwin made it very clear that he was talking about the tribe and not the species, and he stressed that he saw the members of the tribe as being interrelated. So the force at work was certainly more akin to a kind of family selection than to outright group (meaning species) selection. Moreover, we have seen how intelligence is important for the sophisticated morality that Darwin thought we humans have.

But as man gradually advanced in intellectual power, and was enabled to trace the more remote consequences of his actions; as he acquired sufficient knowledge to reject baneful customs and superstitions; as he regarded more and more, not only the welfare, but the happiness of his fellow-men; as from habit, following on beneficial experience, instruction and example, his sympathies became more tender and widely diffused, so as to extend to the men of all races, to the imbecile, maimed, and other useless members of society, and finally to the lower animals, – so would the standard of his morality rise higher and higher.

(p. 103)

Morality

Although intelligence itself is produced by (individual) selection, it does force a move up and out of the biological. "With civilised nations, as far as an advanced standard of morality, and an increased number of fairly well-endowed men are concerned, natural selection apparently effects but little; though the fundamental social instincts were originally thus gained" (p. 173). In fact, Darwin worried rather that we Western nations are now acting against our biological self-interest, given the extent to which our medical powers keep alive and fertile those whom natural selection would hitherto have eliminated – not to mention the depressing effects on reproduction of civilization itself. "The careless, squalid, unaspiring Irishman multiplies like rabbits: the frugal, foreseeing, self-respecting, ambitious Scot, stern in his morality, spiritual in his faith, sagacious and disciplined in his intelligence, passes his best years in struggle and in celibacy, marries late, and leaves few behind him" (p. 174, quoting W. Greg). Fortunately: "There are, however, some checks to this downward tendency. We have seen that the intemperate suffer from a high rate of mortality, and the extremely profligate leave few offspring" (pp. 174–5). The educated and civilized do a better job with their children. In other words, fewer children born may not equal fewer children raised to maturity. Biologists refer to this as the difference between r and K selection. If conditions are variable, then it may be better to have more offspring even if you cannot always do a good job of rearing, because in good times you will raise many – r selection. If conditions are stable, then it pays to have a fairly constant number below the absolute maximum and raise all or most – K selection. In our discussion of the origins of religion, we shall meet people who argue that humans obey this rule fairly exactly.

The biology of morality

Moving on to the present: what can we say today about the biology of morality? One has to cut through the thickets a little at this point, because there is no one, definitively accepted position. However, all would agree that Darwin's thinking holds up surprisingly well. Certainly, in the light of today's work on sociobiology and the explanations of the origins of altruism, no one would doubt that a general sentiment of benevolence is produced and maintained by natural selection. There is still debate over the particular way or ways in which this actually occurs. Some believe that one must bite the bullet and appeal to some kind of group-selection hypothesis. Philosopher Elliott Sober and biologist David Sloan Wilson write:

> Human groups are like single genomes, which achieve their unity by being organized to prevent subversion from within as much as possible. It is a great irony that the language of human social control – sheriffs, police, parliaments, rules that enforce fairness, etc. – has been borrowed to describe the social behavior of genes, without the reciprocal conclusion being drawn that human social groups can be like genomes. When we are relieved of the burden of having to explain human behavior without resort to group selection, these connections become obvious and the groupish side of human nature can be taken at face value. Group selection has not been the only force in human evolution – the groupish side is only one side – but in all likelihood it has been a tremendously powerful force.
>
> (Sober and Wilson 1997, 332–3)

Others (most) think that individual selection can still probably do it all. Obviously kin selection is going to be important. In fact, David Hume was ahead of the Darwinians on this point, noting that our moral sentiments tend to track our blood relationships. Charity, as they say, begins at home. "A man naturally loves his children better than his nephews, his nephews better than his cousins, his cousins better than strangers, where everything else is equal. Hence arise our companion measures of duty, in preferring one to the other. Our sense of duty always follows the common and natural course of our passions" (Hume 1965, 483–4). But obviously kin selection cannot do it all. We do have moral sentiments towards those with whom we have no blood ties. Here some version of reciprocal altruism is needed, as Darwin noted.

Basically, you need a situation where it is in your interests to be generous and giving to others, because in turn they will be generous and giving to you, and benefits received (or at least benefits available) outweigh benefits given (or benefits that one is prepared to give). In other words, you need a situation where it is worth having a social contract – you put in your share, but at the same time you are free to take out as needed. The Harvard philosopher John Rawls (1971) proposed one of the best-known social contract theories in recent years. He asked what kind of society would best benefit its members, and he responded that it would be one that was just, where justice was spelt out in terms of fairness. But what is it to be fair? It is not necessarily to give everyone equal amounts. Rather it is to have a society where, although people are different, everyone gets the most that they possibly can, while weighing the needs of others. If one knew that (for example) one was going to be female, talented, and healthy, then one might, out of selfishness, demand of society that it maximally reward the female, talented, and healthy. But suppose that you do not know what you are going to be – you are behind a "veil of ignorance." You might end up as male, thick, and sickly, in which case, in the society we have imagined, you would lose out. So you want a society that looks after the unfortunates as well as the fortunates. This does not mean that everyone gets the same. If the only way you can get the best people to become doctors is by paying them twice as much as professors, since everyone gains by having good doctors that kind of inequality is allowable for the benefit of all. Behind the veil of ignorance, even though you know you might not be a highly paid doctor, you would opt for highly paid doctors, because you know that you would benefit from good quality healthcare.

The problem with social contract theories has always been that they are surely fictional. It was not the case that a group of tribal elders sat down and dictated how things would be, and, even if they did, there is no reason why people should think the results right and proper. But as Rawls himself points out, an evolutionary account seems to slip in here rather nicely.

In arguing for the greater stability of the principles of justice I have assumed that certain psychological laws are true, or approximately so. I shall not pursue the question of stability beyond this point. We may note however that one might ask how it is that human beings have acquired a nature described by these psychological principles. The theory of evolution would suggest that it is the outcome of natural selection; the capacity for a sense of justice and the moral feelings is an adaptation of mankind to its place in nature. As ethologists maintain, the behavior patterns of a species, and the psychological mechanisms of their acquisition, are just as much its characteristics as are the distinctive features of its bodily structures; and these patterns of behavior have an evolution exactly as organs and bones do. It seems clear that for members of a species which lives in stable social groups, the ability to comply with fair cooperative arrangements and to develop the sentiments necessary to support them is highly advantageous, especially when individuals have a long life and are dependent on one another. These conditions guarantee innumerable occasions when mutual justice consistently adhered to is beneficial to all parties.

(Rawls 1971, 502–3)

Much work recently has been directed to showing how game theory can be illuminating in these circumstances. Philosopher Brian Skyrms (1998) has been leading the way, showing that some kind of reciprocation can be an evolutionarily stable strategy (ESS). Suppose, to take a simple example, there are a hundred units of some resource and two individuals are competing for this resource. Suppose, also, that if together they demand more than a hundred units neither gets anything: in other words, if one competitor demands seventy units, the other can only demand thirty units or it will lose out completely. Why then should one not simply demand more than one's share, knowing that others will simply have to go along with this?

Suppose we put this game in the evolution context that we have developed. What pure strategies are evolutionarily stable? There is exactly one: Demand half! First, it is evolutionarily stable. In a population in which all demand half, all get half. A mutant who demanded more of the natives would get nothing; a mutant who demanded less would get less. Next, no other pure strategy is evolutionarily stable. Assume a population of players who demand x, where $x < \frac{1}{2}$. Mutants who demand $\frac{1}{2}$ of the natives will get $\frac{1}{2}$ and can invade. Next consider a population of players who demand x, where $x > \frac{1}{2}$. They get nothing. Mutants who demand y, where $0 < y < (1-x)$ of the natives will get y and can invade. So can mutants who demand a $\frac{1}{2}$, for although they get nothing in encounters with natives, they get $\frac{1}{2}$ in encounters with each other. Likewise, they can invade a population of natives who all demand 1. Here the symmetry requirement imposed by the evolutionary setting by itself selects a unique equilibrium from the infinite number of strict Nash equilibria of the two-person game. The "Darwinian Veil of Ignorance" gives an egalitarian solution.

(Skyrms 2002, 276)

(A "Nash equilibrium" is a situation where you cannot better your position if others do not change their positions. Suppose there are 100 units of some desirable thing up for grabs, and that if more than a 100 units are demanded no one gets anything. If player A demands 70 units, you as player B cannot demand more than 30 units. If the units are divisible, then if A has an infinite number of options,

Morality

and there will be a corresponding infinite number of Nash equilibria. Skyrms's point is that only 50:50 is both a Nash equilibrium and an ESS.)

The meaning of morality

Note, as Darwin realized fully, that simply getting the biology to work does not give you morality. Ants cooperate much better than humans, but they are not moral beings. Morality demands some kind of ability to reflect on feelings and actions. The moral person is the person who knows the difference between right and wrong and who acts accordingly. You might want to ask why it is that we do have such an elaborate system as morality, especially when sometimes – too often perhaps – people do not do what they ought to do, they are selfish. Would it not be better if we were all like the ants and did the right thing instinctively all of the time? Perhaps so, but even if one could get human brains wired up so that we always do the right thing, there are good reasons why it might not be such a good idea after all. This is tied in with the fact that (compared to the ants) we have only a few offspring but look after them. (Compared to ants, we humans are K-selected, they are r-selected.) Ants act instinctively. Normally this works perfectly, but if something goes badly wrong then usually they cannot adjust and they change their behavior accordingly. For instance, if there is a rainfall and this destroys the chemical trails that they use, then many ants are simply going to be isolated from the main group and die.

The well-known American philosopher Daniel Dennett tells the amusing but revealing story of the wasp, *Sphex*. The wasp paralyses a cricket, brings it to its den, and then drags the body in for its offspring to devour.

> The wasp's routine is to bring the paralyzed cricket to the burrow, leave it on the threshold, go inside to see that all is well, emerge, and then drag the cricket in. If the cricket is moved a few inches away while the wasp is inside making her preliminary inspection, the wasp on emerging from the burrow, will bring the cricket back to the threshold but not inside, and will then repeat the preparatory procedure of entering the burrow to see that everything is all right.
>
> (Dennett 1984, 11)

Such behavior can go on without end. "The wasp never thinks of pulling the cricket straight in. On one occasion the procedure was repeated forty times, always with the same result".

Humans cannot behave like ants and wasps. They have so few off-spring that they need the flexibility to respond to challenges and changes. This is where intelligence comes in, and it certainly plays a role in morality. For instance, in calculating whether one should give in certain circumstances, one might be more inclined to give to some-one who is handicapped than to someone who is healthy but does not want to make any effort.

Why bother with morality?

You might be spurred to put the question another way and ask why you bother at all with morality. Why not just be a calculating machine and decide whether or not actions are in your interests: if they are, go forward, and if they are not, go back. Well, of course, this is often what we do in real life. Many of our interactions with our fellow humans are less matters of morality and more of calculation. I need something and so I go to the store to purchase it; I pay for the item and both sides are satisfied. The problem with this is that it all takes time. And time itself is money, especially in biology. Making calculations all of the time would often mean that things would not get done quickly enough or not get done at all. So morality, in a way, is a quick and dirty solution to life's challenges. I see someone in the road about to be knocked down by a truck. If I spend time calculating whether I should warn them or push them out of the way, it will be far too late. If I shout out or dash over and push them, it may have costs, but the person in danger is better off for my quick actions and con-versely I am better off relatedly the next time I need instant help.

Morality, therefore, is a kind of middle path between being totally controlled by the genes and being super calculating machines that act only in their own best interests. This, incidentally, at once offers an escape from the fear that any kind of biological approach to humankind at once commits us to some kind of "biological deter-minism," where humans are seen simply as robots – a popular image is of human marionettes on strings controlled by the double helix. At one level, the biologist obviously does think that we are determined, by our biology and other factors. This is the assumption one makes when and if one tries to expiate human nature and action scientifically. At another level, however, one has a dimension of freedom – certainly a dimension of freedom over and above the ants, which truly are genetically determined. We are like the vehicles put on other planets – Mars Rover, most notably. They are determined in the sense that they run according to laws, but they have freedom in the sense that they can respond to changes and challenges like rocks in their path: they do not have to wait for instructions from earth. Thanks to morality, we too have a dimension of freedom. We are faced with a challenge, and then we must decide what to do. We are not free to choose our morality, but no one other than philosophers at their most extreme and unconvincing, like Jean-Paul Sartre (1948), have ever thought that we were. But we are not locked into one and only one course of action by our biology.

Normative ethics and metaethics

To conclude this discussion, let us move now from the more scientific aspects of ethics to the more philosophical. What kind of normative or substantive ethics are we getting here? We have already given strong hints about the answer to this question, particularly in discussing Rawls's theory of justice. It is going to be heavily weighted towards reciprocation: do as you would be done by. I suspect that it is a mistake to try to pin things down much more than this. What one expects is a commonsense morality, the kind of morality that is basically covered by all of the great systems, religious and secular. Help others and expect help in return. And in line with Hume's sentiment, the closer you are to home the greater the moral imperative. You have obligations to your own children that you do not have to the children of others. Anything more by way of refinements is going to be a matter of culture as much as (or more than) biology – although notice that if the culture gets too far from the biology, then at most lip service gets paid to it. Jesus may have exhorted us to give all that we have to the poor, but generally even the best Christians take this with a pinch of salt.

Is it not the case, however, that, knowing that morality is controlled by the genes, we could, as it were, do an end run around them? If we wanted to revise the basic principles, we could simply plot to work around biology. Richard Rorty, for all of his supposed enthusiasm for Darwinism, seems to think that this is a possibility.

> Knowing more details about how the diodes in your computer are laid out may, in some cases, help you decide what software to buy. But now imagine that we are debating the merits of a proposed change in what we tell our kids about right and wrong. The neurobiologists intervene, explaining that the novel moral code will not compute. We have, they tell us, run up against hard-wired limits: our neural layout permits us to formulate and commend the proposed change, but makes it impossible for us to adopt it. Surely our reaction to such an intervention would be, "You might be right, but let's try adopting it and see what happens; maybe our brains are a bit more flexible than you think." It is hard to imagine our taking the biologists' word as final on such matters, for that would amount to giving them a veto over utopian moral initiatives.
>
> (Rorty 2006)

Fortunately, or unfortunately, that is precisely what is being claimed here. Biology does matter. Jesus can say what he likes. We may agree with him that we should give all to the poor, but – as the Christian knows full well – that is not what we are going to do, and by and large we are not going to feel that guilty. Biology does have "a veto over utopian moral initiatives." We might get into some kind of Dr. Frankenstein plan for modifying the brain, surgically or through drugs – in fact, this is what we do already today with some kinds of evildoers, as the movie *One Flew Over the Cuckoo's Nest* illustrates rather horribly – but modify we must, or recognize the limits. Because this point is so obvious, one suspects that Rorty's plaintive cry truly is more than just one about the power of biology at the normative level. He worries also about metaethical issues. Is biology now going to justify what we do? Have disinterested moral norms now become simply a matter of the genes as chosen by natural selection? So let us ask now: What of foundations? What of the metaethical support for the kind of substantive ethics that is yielded by Darwinian biology?

My suspicion, one that I think is shared by Hume, is that it is probably a mistake to look too closely for support of this kind. The best that we can do is to say that this is the way that we think and behave. We must leave it at that. There is no support over and above this. At least, there is no support unless you want to start invoking God; in that case, presumably, you are going to have some kind of natural-law theory such as is favored by Roman Catholics (Ruse 2001). God has made the world in the way he has, and part of this making is to fill us with sentiments about the ways in which we ought to behave. These sentiments are natural in the sense that they accord with our nature, and hence it is right and proper to go with them. Of course, there can be much dispute about what is natural for animals such as humans, but this is another matter. There is really no dispute with the memorable sentiment expressed in the *African Queen*: "Nature, Mr. Allnut, is what we are put in this world to rise above." It is part of our human nature to rise above the gross desires of our other nature, our animal legacy.

What about the obvious objection that, even though we have our moral feelings through evolutionarily acquired adaptations, this is not to deny that there might be independent moral values of some kind? After all, we see the train with our eyes, evolutionarily acquired adaptations, but this is not to deny the reality of the train. However, apart from the difficulty of knowing what these independent moral values might be – other than perhaps the sentiments of the deity, and, as we have just seen, if they are these then perhaps they are not so very independent – we need some assurance that the values are actually being captured by our sentiments or feelings. It does not seem above the realm of possibility that we might have evolved in some very different way. We might (say) have a feeling that we should hate our neighbors but, since they feel the same way, get into a kind of reciprocation situation very much as if we loved them. After all, this kind of thing worked pretty well between America and Russia during the Cold War. But this seems to suggest that we might have these independent moral values but do exactly the opposite to what they demand. Unless some account is given of them, as we can give an account of trains, I am not sure that we need worry about them at all.

In a similar vein, I think that this spells failure for some kind of neo-Kantian position where what we do and think morally is given support by being a necessary condition of our being a social animal. At least, as Darwin himself has pointed out, it does seem that if our biology had taken a different path, we might have a very different kind of morality – one where mothers have obligations to kill daughters, and sisters have obligations to kill brothers. It is better, really, to stay with some kind of neo-Humean position where morality is all a matter of the way we are and feel. Although the urge to have something more in some way than mere feeling does point to one interesting fact. Even though morality is a matter of feeling, part of this feeling is that it is not just feeling! In the words of the philosopher John Mackie (1977), we "objectify" moral sentiments. Morality may be subjective, but we feel that it is objective. And for a very good (Darwinian) reason. If we thought that morality was just a matter of feeling, before long it would lose its authority over us and break down. It has to have some hold that makes us want to take it seriously, this hold being that we think morality is something external to us all. You ought not kill, not because it is against your feelings, but because it really and truly is wrong. The Darwinian position is a bit like that of the emotivists, in the sense that it takes morality to be a matter of feeling. But it differs from the emotivists in realizing that the meaning of morality is that it is not just feeling, but something more. It succeeds by taking us in.

Hence, although morality to the Darwinian is simply an adaptation, it is an adaptation of a very particular kind. And that is a good point on which to end the discussion.

10
Religious Belief

Darwinian evolutionary theory is the child of Christianity. As in so many relationships, one sees friendship, one sees tension, and every now and then one sees glimpses of each in the other in the most unexpected of places. Let us begin with Charles Darwin's own religious views, then move on to the relationship between Darwinism and Christianity, and conclude with looking at some of the historical and contemporary issues that have emerged from the *Origin* and its ideas. Here I want to see how Darwin's ideas impinge on issues in the philosophy of religion. I will leave for a separate chapter discussion of the relevance of evolutionary ideas in understanding the origins of religion.

Darwin and religion

As a child, Darwin was raised an Anglican, a member of the Church of England. He admitted that as a youth he did not think very hard on these matters, but said that he had no great difficulty with the Thirty-Nine Articles (although he was not sure that he fully understood them) or with the general truths of Christianity. Indeed, he remarked that on the *Beagle* voyage the sailors used to laugh at him for taking so literal an approach to the Bible. Darwin was trained at Cambridge in a Christian atmosphere, the dominant works being those of Archdeacon William Paley on revealed religion (*Evidences of Christianity*) and on ethical behavior (*The Principles of Moral and Political Philosophy*). Darwin also read, and was very much taken with, Paley's book on reasons for belief (*Natural Theology*). Darwin's

Christian beliefs started to fade away on the *Beagle* voyage. Particularly, he began to doubt the veracity of the biblical miracles, considered as violations of the laws of nature, as well as to wonder why Christianity should be privileged over other world religions, like Hinduism. One will perhaps wonder why the feeling that the miracles might not be literally true should be so important – after all, a good many Christians today prefer to think of them as events of great significance and downplay their supernatural qualities. To understand Darwin here one must understand Paley. The *Evidences of Christianity* is no work of a Kierkegaardian or (Karl) Barthian strain, in which only the route to the Creator is (unjustifiable) faith; it is rather an attempt to prove that it is reasonable to believe the gospels. In particular, Paley argued that the miracles must be true because the disciples were prepared to die for their belief that Jesus was divine. Had the miracles not occurred as described in the Bible, the disciples would have pulled back. Conversely, the truth of the miracles makes plausible the truth of Jesus' divinity. For Darwin, when the miracles went, the divine status of Jesus went too.

In his *Autobiography*, written late in life, Darwin presented himself as doubting the miracles on literary and scriptural grounds – the Tower of Babel is just not reasonable – but undoubtedly by the time of the *Beagle* voyage the influence of Lyell was starting to kick in. The *Principles of Geology* went strongly against a miracle-dominated world, as well as demanding a non-literal account of earth's history. As Lyell's most enthusiastic student, Darwin was bound to be influenced. And then on top of this, at some point now or later, the eschatological (end time) issues were becoming important. Darwin was one of many Victorian intellectuals who worried themselves sick about eternal punishment. It just did not seem reasonable that good people who did not believe should be condemned to everlasting hell and torment. His father was a nonbeliever and yet to Darwin he was the epitome of the good man.

None of this should be taken to imply that Darwin at once became an atheist. He never felt an atheist or argued for atheism. Later in life, he did become an agnostic – a skeptic – like just about everyone else in his scientific set. But, by the mid-1830s, Darwin is best described as a deist, one who thinks of God as unmoved mover, who, having set the world in motion, now lets all unfurl according to unbroken law. A deist is distinguished from a theist, the latter (a term usually reserved for Christians, Jews, and Muslims) being someone who believes in a God who intervenes in his creation, significantly (in the case of Christians) by sending his only-begotten son to die for our sins. Lyell was a deist, and so were the Wedgwoods. They were Unitarians, meaning that they denied the existence of the Trinity (and hence thought that Jesus was just a good man), and wanted no part of a God who keeps interfering down here. This was a common position among people who had done well out of the Industrial Revolution. A popular metaphor depicted the deistic God as a god of machinery who set things running, as opposed to the theistic God who has to work by hand. Charles Babbage (1837), inventor of the computer and a good friend of Charles's older brother Erasmus, made much of this picture.

Deism seems to have been Darwin's position right through the writing of the *Origin*. Just after its appearance, he admitted to Asa Gray, a committed, evangelical Presbyterian, that he could not see God in the details but he still thought that there was something out there behind everything.

> I see no necessity in the belief that the eye was expressly designed. On the other hand I cannot anyhow be contented to view this wonderful universe & especially the nature of man, & to conclude that everything is the result of brute force. I am inclined to look at everything as resulting from designed laws, with the details, whether good or bad, left to the working out of what we may call chance.
> (Letter to Asa Gray, May 22, 1860; Darwin 1985–, 8: 224)

Nevertheless, even if Darwin had rejected the Christian God as a matter of belief, the Christian training was still very significant. If it was deism that made Darwin an evolutionist, it was Christian theism that made him a Darwinian, meaning one who thought that the mechanism was natural selection.

Christian influences

How can one justify this claim about the connection between Christianity and Darwinism? By recognizing Paley's Christianized natural theology as the very foundation and basis of Charles Darwin's creed that the key to understanding the living world is adaptation – or, as he sometimes called it, contrivance (Ruse 2003). Organisms are not just thrown together higgledy-piggledy, but are highly sophisticated survival and breeding machines. They have features that help them in this endeavor – features like eyes and noses and teeth and penises and vaginas, features like leaves and trunks and flowers and roots and pistils and stamens and pollen. In the world of natural theology, these adaptations are the key factor in the proof of God's existence. The eye is like a telescope; telescopes have telescope makers; therefore eyes have eye makers – what I have called the Great Optician in the Sky. Darwin accepted the premise entirely. Organic parts are as if designed – what Richard Dawkins (1983) has described as "organized complexity." What Darwin sought was not a refutation of the conclusion – right through the *Origin* he went on believing in the Great Optician – but some way in which he could account for adaptation through law. This was why selection proved so important to him, because he could see that it did lead, not just to change, but to change in the direction of features that appeared to be designed. Completing the story down to the present, it was this centrality of design explanation (or, if you prefer, design-like explanation) that gave Darwin's theory a logical structure different from that of the physical sciences. Darwinians are with Kant in thinking that final-cause explanations are meaningful and essential.

There was still the matter of accounting for a natural form of selection, and for this Darwin needed the struggle for existence. Note that this move, too, required Darwin to turn to Anglican theology; but in this case he did reject the conclusion – namely, that all significant and long-lasting change is impossible. Malthus introduced his ratios to speak to the issue of how God gets us to work, while we are down here on earth. The answer is that if we did not we would starve, because the population pressures are such that soon there would not be food for anyone. So we labor, even though most will suffer and unlimited change is impossible. Darwin took on completely the Malthusian arithmetic and most probably agreed also with the intent. As a child of the Industrial Revolution, he knew all about the need to work, and moreover would have agreed that unrestricted handouts were not the solution to life's problems. Apart from anything else, without the harsh laws of population, no one would take the paltry wages that were all industrialists like the Wedgwoods could pay if they were to build their empires. The Darwins and the Wedgwoods were strongly against slavery, but they had no problems with depressed pay for men and women who were unable to escape the misfortunes of their low birth status.

If you combine this with the way in which, after the *Origin*, evolution took on the role of a secular religion – a story of origins, progress, humans at the top – you start to see why it is that I claim that Darwinism is the offspring of Christianity. A bastard offspring, if you insist, fathered by the Enlightenment, but an offspring nevertheless.

Genesis

It is hardly surprising that after the *Origin* many found Darwin's theory to be unacceptable. What is surprising is how many found it acceptable and how many of these were religious people (Roberts 1988; Numbers 1998). Almost everyone wanted special interventions for human souls, and few were entirely satisfied with natural selection, but then few agnostics were entirely satisfied with natural selection either. It is hard to quantify these things, but by the mid-1870s the expectation would have been that an Anglican or a Presbyterian or member of a central dissenting (nonconformist) church in Britain and America (and their equivalent in Europe) would accept some form of evolution. At first, many Catholics also accepted evolution, although in both Britain and America Catholics tended to be in the lower segments of society and hence not really interested at all in the religion and science interaction. Unfortunately, it was also around this time that the Catholic Church generally became more conservative – in major part because of the papacy's political problems with the unification of Italy – and acceptance of scientific advances dropped accordingly.

The Reverend Henry Ward Beecher, brother of the novelist Harriet Beecher Stowe, charismatic preacher and adulterer, spoke for many:

> If single acts would evince design, how much more a vast universe, that by inherent laws gradually builded itself, and then created its own plants and animals, a universe so adjusted that it left by the way the poorest things, and steadily wrought toward more complex, ingenious, and beautiful results! . . . Who designed this mighty machine, created matter, gave to it

its laws, and impressed upon it that tendency which has brought forth the almost infinite results on the globe, and wrought them into a perfect system? Design by wholesale is grander than design by retail.

(Beecher 1885, 113)

To the conservative Anglican Edward Pusey, in response to a letter about a proposed honorary degree for Darwin from Oxford University, John Henry Newman, former Anglican, later to be a prince cardinal in the Church of Rome, mused: "Is this [Darwin's theory] against the distinct teaching of the inspired text? If it is, then he advocates an Antichristian theory. For myself, speaking under correction, I don't see that it does – contradict it" (Letter of June 5, 1870; Newman 1973, 137).

If Christians did accept Darwin's ideas, why then was there such a massive row, and why even today is there opposition to Darwinism by religious people? The most obvious answer is that Darwinism is in theory opposed to Christianity. And the most obvious point of conflict is the clash between the story of creation as given by Genesis, the opening chapter of the book that Christians think infallible, and the story of creation as given by Darwinism. The one claims that all was done in six days a short length of time ago (within the last 10,000 years) through miracle, and the other claims that it took millions of years and was entirely lawbound.

Things are a bit more complex than this. At least since the time of St. Augustine, around 400 AD, it has never been part of the Christian tradition that one must necessarily read the Bible in a totally literal way (McMullin 1986). Augustine himself had, as a young man, been a Manichean, a sect that did not accept the Old Testament as canonical, so he knew all of problems – the two not entirely consistent creation stories of Genesis, for instance – and he insisted that, although the presumption must be that the Bible is literally true, if common sense and science dictate against this, then one must interpret it in an allegorical fashion. The Bible was written for primitive, nomadic folk, not for sophisticated members of the Roman Empire. Pertinent, in the particular case of origins, was Augustine's belief that God stands outside time. This means that, for God, the thought of creation, the act of creation, and the completion of creation are as one. Augustine was certainly not an evolutionist as we understand the term, but he did believe that God created the seeds of life – the potentiality – that then unfurled. The metaphysics was one that lends itself to evolution.

The coming of the Protestants generated enthusiasm for literal readings of the Bible. For mainstream Protestants – Luther and Calvin – the Bible takes the place that tradition and the church have for Catholics. (Some members of the more radical Reformation movements, including in the Anglo-Saxon world the Quakers, put more emphasis on the Holy Ghost, "that of God in every person.") But even the great reformers did not always insist on literal readings. Luther, whose theology was centered on justification by faith – God's grace is the beginning and end of salvation – hated the Epistle of James because it says explicitly that good works are important. Calvin, like Luther a follower of Augustine, spoke of God's need to accommodate his language to the unsophisticated, and this apparently meant that non-literal readings of the Bible were permissible, as necessary. With the coming of Copernicanism, non-literal interpretation was essential if one was to make sense of the sun stopping for Joshua as he fought the Amorites.

Although, at the time that Darwin wrote, there were some literalists in Britain, they were considered outside the fold. General opinion, even among catastrophists, was that the earth is very old, and that the six days of creation are either long periods of time or that the Bible does not mention long periods of time that occurred between the (literal) days of creation. The supposed worldwide flood, associated in the Bible with Noah, was also now being read as something more limited. Remember that Darwin's shipmates laughed at him for being so literal. The big issue for Christians (Protestants especially) on the revealed religion side of things was the perceived clash between Providence and progress. Progress implies that change is possible and that we humans unaided are the driving force. You can have better education, cleaner water, better working conditions, proper medical care, just so long as you work hard to get these things. For better or for worse, evolution was seen as allied with these notions of upward improvement, coming from within (Ruse 2005). Providence goes the other way. Nothing, absolutely nothing, that we do comes from our own efforts or merit. For the Protestant (especially the Protestant of the pre-Darwinian era) all ultimately is a matter of God's grace – we are as nothing without his help. As Augustus Montague Toplady's famous hymn "Rock of ages" (1775) puts it:

Not the labor of my hands
Can fulfill Thy law's demands;
Could my zeal no respite know,
Could my tears forever flow,
All for sin could not atone;
Thou must save, and Thou alone.

American literalism

If, then, Christianity did have a tradition of biblical interpretation, including allegorical and metaphorical readings of passages that seem to conflict
with science, why was it that after the *Origin* so many people – especially so many people in America – wanted to reject Darwinism on
the grounds that it conflicted with Genesis? Why was it that in 1925
in the state of Tennessee, a young schoolteacher – John Thomas Scopes
– was prosecuted for teaching evolution (and found guilty and fined)?
The real answer is bound up with American social history. Modern
America was ever a religious nation, from its founding by the Puritans
fleeing Europe to worship in their own fashion. After the Revolution,
Protestant religion filled the gaps left by the British defeat, and by
the middle of the nineteenth century an indigenous form of Christianity was forming – one that put heavy emphasis on the Bible as a
guide to life in the New World (Noll 2002).

The Civil War accentuated stark differences in the nation. In the
North, this was the time of great industrial moves forward and of the
embracing of modern ideas, including those in religion. The so-called
"social gospel" became dominant, with heavy emphasis on good
works and corresponding de-emphasis on embarrassing tales from
scripture. In the South, and increasingly in the West as the railroads
opened up new horizons, a biblically based religion intensified, one
that shows how God inflicts troubles on his most favored peoples
and that gives hope for the future, especially if there is proper interpretation of the last, apocalyptic book of the Bible, Revelation. This
kind of theology – which, I emphasize, is a newly made, idiosyncratic
form of nineteenth-century American Protestantism – puts a heavy
emphasis on literalism, especially of the early chapters of Genesis. It
sets itself against geology and against biology, and the universality of
the flood is a lynchpin because of its symmetrical correspondence to

the expected Armageddon. Evolution is seen as a kind of litmus test for everything that is wrong with modern (read Northern) society, and must be opposed (Numbers 1992).

The Scopes "Monkey Trial" (1925) was as much a protest against modernism as it was a debate about history (Larson 1997). Indeed, as a matter of fact, the man leading the prosecution in the trial – William Jennings Bryan, three-time presidential candidate, former secretary of state under Woodrow Wilson – did not subscribe at all to a short lifespan for the earth. "But, beloved, be not ignorant of this one thing, that one day is with the Lord as a thousand years, and a thousand years as one day" (2 Peter 3: 8). The strict limitation of creation to a literal six days 6,000 years ago was always an odd belief of that nineteenth-century sect the Seventh-day Adventists, and did not become popularized until after the Second World War. But then and now, time was never the really big issue. Literalists, once known as Fundamentalists and more recently as Creationists, are opposed to modern society and its customs as much as to modern science and its claims. "It is rather obvious that the modern opposition to capital punishment for murder and the general tendency toward leniency in punishment for other serious crimes are directly related to the strong emphasis on evolutionary determinism that has characterized much of this century." Apparently, the "notorious Darwinian philosopher Michael Ruse," well known as an "atheistic humanist," has been a key player in America's decline into the moral abyss (Morris 1989, 148).

Summing up. There is no tradition in Christianity that mandates the literal reading of the Bible and much to suggest that the Christian is obliged to think allegorically or metaphorically when science so dictates. In America to this day, Creationist opposition to evolution is strong, but it is as much cultural and social as it is theological. Genesis need not be in opposition to Darwinism.

Progress (again)

Yet surely allegory and metaphor can only go so far? You might argue about the meaning of the death of Jesus on the cross. Was he a sacrificial lamb and if so to whom was he being sacrificed? Was his death a payment for our sins and if so to whom was the payment being made? What you cannot deny (and still claim to be a Christian) is that Jesus did die on the cross and because of this we humans benefited. It is the same with Genesis. You can argue all you like about the veracity of Adam and Eve and so forth, but there is a bottom line to the argument and to breaking from the literal word of the text. You have to accept that God is Creator and Lord. You have to accept that God made us humans and that a reason why we have a special place in his heart – a place so big that he died on the cross to make possible our eternal salvation – is that we are made in his image. You can argue at length about what it is to be made in God's image. But whatever it is – and presumably intelligence and a moral sense are bound up here – we simply cannot be just contingent. If the Bible tells us anything, it is that we humans cannot be a mistake or an afterthought. We may not be the only reason for the creation, but we are a central part of the divine plan.

This means that, for the Christian, evolution had to produce humans. I do not mean that evolution had to produce humans exactly as we are – ten fingers, thirty-two teeth, foreskins – but it had to produce intelligent beings with a sense of morality and so forth. Perhaps it did not have to happen on planet earth, but it had to happen somewhere in the universe. Somewhere, some time, something like us had to occur. Does this not bring us right back to the issue of progress and of absolute value? Without progress and absolute value, the Christian claims are negated. Actually, not quite! At the theological level, many Christians would argue that the processes and our understanding of them are irrelevant. St. Augustine's theology, which places God outside time, can carry the burden. Since, at the creation of the world, seeds of potentiality were already there, life was going to unfurl whatever the process, and humans would arrive in due time. Don't bother about details. We were supposed to arrive on earth and we did.

You might think that there is something a bit deterministic about this, and you are right: there is something a bit deterministic about Augustine's thinking altogether. It was no wonder that Calvin, the master of predestination, liked him so much. So, what if you want to be a little less upfront theological about these matters? Do note that the worry that we need progress and absolute value are not the big issues here. We can read in progress and absolute value ourselves. It is just that we cannot claim that they are necessarily delivered

Religious Belief

by Darwinian evolution. But that is no problem. The nonbeliever like me can think that humans are better than all other beings. It is just that I read this into nature. I do not find it in nature. Having said this, however, there is still work to be done. You have still got to get the production of humans, valuable or not. Here it seems we have to go back to arms races or scrambling into niches or whatever.

What if you are Gould and do not care much for any of this? "In an entirely literal sense, we owe our existence, as large and reasoning mammals, to our lucky stars" (Gould 1989, 318). You could argue that Gould was not keen on the moves that lead progressively to humans – arms races, for instance – precisely because he was not keen on hard-line Darwinism, and these moves are very much hard-line Darwinian. But while that may be so, I am not sure that all hard-line Darwinians necessarily think that they guarantee the appearance of humans. There is another tack that you can take, one that Gould himself endorsed because (perhaps surprisingly), despite his sneers at progress, he thought that human-like beings were prob-able somewhere in the universe – and they may have appeared more than once! "I can present a good argument from 'evolutionary theory' against the repetition of anything like a human body elsewhere; I cannot extend it to the general proposition that intelligence in some form might pervade the universe" (quoted by Dick 1996, 395). Indeed, in an argument predating that of Conway-Morris, Gould suggested that culture might be a niche that has been invaded more than once.

If he was not going to buy progressivist processes, what force did push things along for Gould? What perhaps even made it seem as though life was moving upward toward intelligence? Gould (1996) argued that a kind of directionality comes about through random processes and nature's constraints. Life is a bit open-ended. It started simple. It cannot get simpler. It can get more complex. Not through any direc-tion, but because this is the way that things are. It is rather like the drunkard and the sidewalk. On one side, the sidewalk is bounded by a wall, and on the other side lies the gutter. Eventually, the drunkard will end in the gutter, not because he intends this but because he cannot walk through the wall. His random staggering will eventually lead to the gutter. So it is with evolution. There is no progress in nature, but there is direction toward forms like humans because, eventually, ever-searching life will hit the jackpot.

This will probably be enough for the Christian. Yet, is not the Darwinian now left feeling a bit blue, because the answer comes only by jettisoning natural selection as a significant factor? Perhaps so, but interestingly the answer is one that Darwin himself toyed with. In an entry in a private notebook from January 1839 – just months after he first hit on selection and how it works, and twenty years before the *Origin* – he speculated on just the ideas that Gould found attractive:

> The enormous number of animals in the world depends, of their varied structure & complexity.– hence as the forms became complicated, they opened fresh, means of adding to their complexity.– but yet there is no NECESSARY tendency in the simple animals to become complicated although all perhaps will have done so from the new relations caused by the advancing complexity of others.
>
> (E 95; Darwin 1987, 422)

Note the adamant denial that there is any necessity to the directions of nature – the *Naturphilosophen* are the target here. Then Darwin added:

> The Geologico-geographico changes must tend sometimes to augment & sometimes to simplify structures:– Without enormous complexity, it is impossible to cover whole surface of world with life.– for otherwise a frost if killing the vegetable in one quarter of the world would kill all of the one herbivorous. & its one carnivorous devourer.;– it is quite clear that a large part of the complexity of structure is adaptation. though perhaps difference between jaguar & tiger may not be so.– Considering the Kingdom of nature as it now is, it would not be possible to simplify the organization of the different beings, (all fishes to the state of the Ammocoetus) Crusacea to – ? &c) without reducing the number of living beings – but there is the strongest possible to increase them, hence the degree of development is either stationary or more probably increases.–
>
> (E 96–7; Darwin 1987, 422)

Darwin's discussion was making a somewhat stronger claim than Gould's because, random factors notwithstanding, Darwin was giving selection some role (although a far from exclusive one) in this process. The important thing is that if God wants humans he will probably get them if he waits long enough.

252   *Religious Belief*

The argument from design

Turn now to natural theology, and focus on the argument from design. There are two questions: "Why was the argument from design so powerful before Darwin?" and "What effect did his theorizing have on it?" As an analogical inference, the design argument has an obvious appeal. Paley was right. The eye is like a telescope. Telescopes have telescope makers. Hence, it is reasonable to conclude that eyes must have eye makers. God! The wonderful thing about the argument is that it does not simply prove God's existence but points to his having some of the attributes that we traditionally associate with the Christian God. As hands and eyes and teeth and so forth are miracles of engineering, and clearly intended for our use and happiness, the Designer must himself be very clever and concerned about human welfare (not to mention the welfare of the other living parts of his creation). Since we humans have the best of all adaptations, we must be the most favored by God.

But there were big problems with the design argument. In his *Dialogues Concerning Natural Religion*, published posthumously in 1779, David Hume had pointed to these. So much of the world seems ill designed – the pains and agonies that all creatures have to endure – that it really does not seem as if a good God could have designed and made things. In any case, why one God only, why not a squad? And is it not reasonable to think that there were many, lesser, botched worlds before ours, and perhaps that there will be better worlds to come after ours? By the time Hume had finished, the design argument seemed smashed to smithereens, and yet, paradoxically, right at the end he too admitted that there might well be something more. "If the proposition before us is that *'the cause or causes of order in the universe probably bear some remote analogy to human intelligence'* . . . what can the most inquisitive, contemplative, and religious man do more than give a plain, philosophical assent to the proposition, as often as it occurs; and believe that the arguments, on which it is established, exceed the objections, which lie against it?" (Hume 1947, 203–4; his italics).

The answer has to be that by the end of the eighteenth century the argument was no longer one of analogy, but had been transformed into something else. It had become what philosophers call an "argument to the best explanation." As Sherlock Holmes put it to his friend Dr. Watson, in *The Sign of Four*: "How often have I told you that when you have eliminated the impossible, whatever remains, however improbable, must be the truth." If you have something that needs explaining, and there is only one possible explanation, then it is reasonable to go with this. The fact is that the living world does seem as if it is designed. Blind law will not account for it. Blind law leads to mess, to decay, to (in the language of the physicists) increase in entropy. Murphy's Law holds: "If something can go wrong, it will," or, in the alternative version: "Bread always falls jammy side down." There must be a reason for the design-like nature of the world, and the only reasonable explanation is an intelligence – God. It is as simple as that.

Now we can see why Darwin had so great an impact. Darwinism and natural theology share an important premise. The living world seems as if designed. The eye is like a telescope. The heart is like a pump. The shell of a tortoise is like a shield. Darwin then went ahead and offered a different explanation. Natural selection on random variation causes adaptation, the design-like nature of the living world. There is no longer a need to invoke an immediate designer. In the words of Richard Dawkins: "although atheism may have been logically tenable before Darwin, Darwin made it possible to be an intellectually fulfilled atheist" (Dawkins 1986, 6). Before the *Origin*, total atheism was simply not reasonable. Notice, however, that this is not to say that one must be an atheist. Only that one can be. As we know, the arguments of the *Origin* did not make Darwin himself an atheist. He thought that there was a designer, albeit one who worked at a distance through unbroken law. Which, of course, means that one does expect some differences between the work of the Paley God and the work of the Darwin God. The Paley God can do the best job possible. The Darwin God can only do the best job possible given unbroken law – that is, given the making of organisms through evolution. We have seen already that the Darwin God expects organisms to look, at times, as though they were made by Rube Goldberg or Heath Robinson, as indeed they do, rather than perfect all of the time, as one would expect from the Paley God.

In short, although Darwin might not have made impossible the God of design, he does rather weaken any argument to God through design. It is more a question of believing in design because of God, than believing in God because of design – a point noted and appreciated by Newman. In 1870, about his new major philosophical work *A Grammar of Assent*, Newman wrote:

> I have not insisted on the argument from *design*, because I am writing for the 19th century, by which, as represented by its philosophers, design is not admitted as proved. And to tell the truth, though I should not wish to preach on the subject, for 40 years I have been unable to see the logical force of the argument myself. I believe in design because I believe in God; not in a God because I see design. . . . Design teaches me power, skill and goodness – not sanctity, not mercy, not a future judgment, which three are of the essence of religion.
>
> (Newman 1973, 97)

By theologians, this kind of approach is known as a "theology of nature" rather than a "natural theology." It is one much favored today by sophisticated Christians (Pannenberg 1993).

Miracles

Darwin's long, slow journey to nonbelief began with his doubts about miracles. Do you have to give up miracles if you are a Darwinian, and does this mean that you can no longer be a Christian? Certainly the issue of miracles is one that impinges on both Darwinism and Christianity, but the topic is a little more complex than one against and one for and never the two shall meet. Start with the fact that all science excludes miracles, if these are understood as violations of nature. Science cannot start unless you assume that the world is governed by unbroken regularities. To assume otherwise is to allow what Alvin Plantinga (1993) has called a "science stopper." Hence, inasmuch as you are a Lyellian, inasmuch as you are a Darwinian, you are committed to the rule of law.

So what about Christians and miracles? Even if one wants to downplay the many miracles that the Catholic Church seems to be prepared to embrace, the bottom line is that one must accept some of the biblical miracles – the raising of Lazarus, the feeding of the five thousand, and above all the resurrection of Jesus on the third day. What is one to do in the face of science? Start with the fact that some miracles – some of the most important miracles – do not at all threaten the scientific picture. The Catholic mass, which involves transubstantiation – the turning of bread and wine into the body and blood of Christ – is an entirely non-natural (meaning outside-the-rule-of-natural-law) phenomenon. Following Aquinas, Catholics believe that it is the substance that is changed and the accidents – those features that we sense, like color and taste – stay the same. Nothing that we do, nothing that we find, has any relevance to this change. The same is true of immortal souls. If as a Christian you believe that an immortal soul is implanted in each human, no amount of study is going to reveal this as a physical fact. You might perhaps worry about how one distinguishes the soul from the mind, where the latter certainly is a natural phenomenon, but that is not a problem for the Darwinian as such – it is a general problem for the Christian.

In dealing with those miracles that do impinge on the natural world, there are two traditional approaches one can take. On the one hand, one can argue with St. Augustine that many miracles – perhaps all miracles – are more a matter of meaning than claims about the physical world, which is never interrupted. The feeding of the five thousand is best understood as the result of people being so moved by Jesus that spontaneously they shared their food with strangers. To think otherwise – to think that Jesus did actually make the food – is to turn him into some kind of high-class caterer, which is hardly fitting for the Son of God. Even the resurrection was no lawbreaking event. It is not a question of what happened to the body of Jesus, but of how, on the third day, the disciples who had been so down-cast by the loss of their leader, were suddenly uplifted and inspired. He really lives – in our hearts, which is what matters. That a psychologist might be able to give an explanation of this is not threatening, but expected.

The other approach is more in the tradition of Aquinas, which holds that nature-violating events did occur. Here the move is to make a traditional distinction between the order of nature and the order of grace. The world works by law normally, but the fact that humans had fallen into sin required the Incarnation and the Atonement and nothing else would do. The God of Christianity holds the world in his hands immanently all of the time. That he would interfere when it was absolutely necessary causes no great theological or ontological shift. Our belief in his interventions is not a scientific claim – it is not reasonable in the sense of science – but it is part of faith, which of course may be backed by reasoned arguments about the theological necessity of God's direct intervention. None of this threatens the general scientific picture, including Darwinism. The whole point is that normally God works by law, and it is only in this very exceptional case that he became directly involved. And there were good reasons.

Intelligent Design

It is well known that today there are those whom the traditional answers do not satisfy. The Intelligent Design (ID) theorists, as they like to call themselves – the Intelligent Design Creationists as their opponents call them – argue that one must allow divine interventions into the normal course of nature, if one is to explain the existence of organisms (Johnson 1991). Regular blind law is not enough. Although the ID theorists pretend at times that this intervener could be natural, in truth they do not think this. The designer is not some bright grad student on Andromeda, manipulating the human race to complete his PhD. The designer is the Christian God; the Logos of the first verse of St. John's Gospel, to adopt a favorite way of expressing it.

Part of the objection to Darwinism is straightforwardly metaphysical and ethical. The leader of the movement, the retired Berkeley lawyer Phillip Johnson (1995), says openly that he dislikes Darwinism because he sees it as part and parcel of a general scientific commitment to naturalism, and he thinks that naturalism leads to morally and socially undesirable consequences – divorce, single parenting, violent teenagers, abortion, gay marriage, cross-dressing, and more. Johnson allows that one can make a distinction between methodological naturalism – where one simply assumes that the world runs according to law – and metaphysical naturalism – where one assumes that science is everything and that there is no God. But he feels that methodological naturalism always collapses into metaphysical naturalism. So all forms of naturalism should be shunned.

Naturalism is a *metaphysical* doctrine, which means simply that it states a particular view of what is ultimately real and unreal. According to naturalism, what is ultimately real is nature, which consists of the fundamental particles that make up what we call matter and energy, together with the natural laws that govern how those particles behave. Nature itself is ultimately all there is, at least as far as we are concerned. To put it another way, nature is a permanently closed system of material causes and effects that can never be influenced by anything outside of itself – by God, for example. To speak of something as "supernatural" is therefore to imply that it is imaginary, and belief in powerful imaginary entities is known as superstition.

(Johnson 1995, 37–8)

The ID theorists back this philosophical discussion with what they intend to be scientific arguments. They argue that there are certain natural phenomena that simply could not be explained by unbroken law, and that a designer had to be involved. In his bestselling book, *Darwin's Black Box*, Michael Behe, a biochemist at Lehigh University, identifies some phenomena as "irreducibly complex."

By irreducibly complex I mean a single system composed of several well-matched, interacting parts that contribute to the basic function, wherein the removal of any one of the parts causes the system to effectively cease functioning. An irreducibly complex system cannot be produced directly (that is, by continuously improving the initial function, which continues to work by the same mechanism) by slight, successive modifications of a precursor system, because any precursor to an irreducibly complex system that is missing a part is by definition nonfunctional.

(Behe 1996, 39)

He adds that any "irreducibly complex biological system, if there is such a thing, would be a powerful challenge to Darwinian evolution. Since natural selection can only choose systems that are already working, then if a biological system cannot be produced gradually it would have to arise as an integrated unit, in one fell swoop, for natural selection to have anything to act on" (p. 39).

Do we have irreducibly complex phenomena in nature? Turn to the microworld of the cell and of the mechanisms we find at that level. Take bacteria, which use a flagellum, driven by a kind of rotary motor, to move around. Every part is incredibly complex, and so are the various parts combined. The external filament of the flagellum (called a "flagellin"), for instance, is a single protein that makes a kind of paddle surface in contact with the liquid during swimming. Near the surface of the cell, just where it is needed, is a thickening, so that the filament can be connected to the rotor drive. This naturally requires a connector, known as a "hook protein." There is no motor in the filament, so the motor has to be somewhere else. "Experiments have demonstrated that it is located at the base of the flagellum, where electron microscopy shows several ring structures occur" (p. 70). All way too complex to have come into being in a gradual fashion. Only a one-step process will do, and this one-step process must involve some sort of designing cause. This is God, even though Behe and his fellows generally try not to bring in the divine quite so brazenly. Their reticence is not theological but political: the US constitution bars the teaching of religion in state-supported schools, and so if ID is to get into science classes it must appear to be religion free.

As you might imagine, the biological community has not been impressed (Pennock 1998; Forrest and Gross 2004). It is denied that there are such things as irreducibly complex organisms or parts. Behe's paradigm case of an irreducibly complex phenomenon is a mousetrap with five parts, all of which must be in place and functioning before the mousetrap works. Many happy hours have been spent by evolutionists building mousetraps with fewer than five parts. You can even get a one-part mousetrap. Not very good, admittedly, but in the land of the blind – in the land of the mousetrapless – the one-eyed man – the one-part mousetrap – is king. The general critique of arguments like Behe's is that you should not presume to say what is or is not irreducibly complex – that is, unproducible by selection – just by looking at the finished product. Take a stone bridge built without cement. Looking at the bridge, you would say it is impossible. As the sides go up and in, they will fall to the ground. But if you first build an embankment and place the stones on this, removing the dirt when you are finished, the bridge will stay in place.

Similarly in the case of nature. Often we find that what went before was not as it is today. There were other structures on which parts were placed: when connections are made, the older structures can be, and often are, removed by selection (Miller 1999). In the case of the motor, evolutionists have found repeated examples of simpler motors that could be built up and refined by selection through the ages. The case of the human or hawk's eye is very much the same: it is said to be too complex to have been produced by selection – there could not be functioning intermediaries. But nature shows that there are and could have been! As people like Richard Dawkins are wont to say, it is always a bad mistake to think that natural selection is not as clever as you are. Too often you will be shown wrong.

There are other problems with ID theory. When did God intervene? Will we see evidence today of his intervening? If it was a long time ago, even before body parts were needed, why were complex features not removed by selection or made inoperative through mutation? Perhaps most worrying about the ID position is that it is a science stopper, and its true purpose is theological. Its supporters want a world of miracles, a world where God is always on call and active, a world where – even though many ID supporters are not biblical literalists – the simple, evangelical-Protestant world picture stays in play. One of the most important ID supporters, mathematician and philosopher of science William Dembski, is open about feeling a strong emotional alignment with traditional "young earth" Creationists.

> Despite my disagreements with . . . young earth creationism, I regard those disagreements as far less serious than my disagreements with the Darwinian materialists. If you will, young earth creationism is at worst off by a few orders of magnitude in misestimating the age of the earth. On the other hand, Darwinism, in ascribing powers of intelligence to blind material forces, is off by infinite orders of magnitude.
>
> (Dembski 2005)

IDT is not science, whatever is being said, but an appeal to ignorance to bolster a religious position. Scientists, including scientists who are Christians, have every right to be suspicious of it. Committing oneself to the rule of law is not blind prejudice, and (as we have seen above in the discussion of miracle) not necessarily to commit oneself to atheism. The simple fact is that, by persisting in trying to find lawbound explanations, again and again scientists have won through. If anything demands the use of our reason, which shows that we are made in the image of God, it is this.

The problem of evil

For me, the biggest problem with ID theory is theological. If God was prepared to get involved to create the irreducibly complex, why was he not also prepared to get involved to prevent the horrendously simple? Many of the worst genetic diseases involve a very minor change at the molecular level. Why did an all-loving, all-powerful God not stop such changes? This is part of the general and

Religious Belief

traditional problem of evil. Let us therefore conclude this chapter by looking at this issue.

Many people, starting with Charles Darwin himself, think that evolution through natural selection exacerbates the problem of evil, forever separating Darwinism from Christianity. Quoting again from Darwin's letter, to his American friend and supporter Asa Gray, written just after the publication of the *Origin*:

> With respect to the theological view of the question; this is always painful to me.– I am bewildered.– I had no intention to write atheistically. But I own that I cannot see, as plainly as others do, & as I sh^d. wish to do, evidence of design & beneficence on all sides of us. There seems to me too much misery in the world. I cannot persuade myself that a beneficent & omnipotent God would have designedly created the Ichneumonidae with the express intention of their feeding within the living bodies of cater-pillars, or that a cat should play with mice. Not believing this, I see no necessity in the belief that the eye was expressly designed.
>
> (Letter to Asa Gray, May 22, 1860; Darwin 1985–, 8: 224)

Today's evolutionists often agree. Even if God does exist, he is certainly nothing like the Christian God: he is unkind, unfair, totally indifferent. In the words of Richard Dawkins:

> If Nature were kind, she would at least make the minor concession of anes-thetizing caterpillars before they are eaten alive from within. But Nature is neither kind nor unkind. She is neither against suffering nor for it. Nature is not interested one way or the other in suffering, unless it affects the survival of DNA. . . . The total amount of suffering per year in the natural world is beyond all decent contemplation.

Dawkins concludes:

> As that unhappy poet A. E. Housman put it:
>
>> For Nature, heartless, witless Nature
>> Will neither know nor care.
>
> DNA neither knows nor cares. DNA just is. And we dance to its music.
>
> (Dawkins 1995, 131–3)

What can be said in response? Let us make the traditional distinc-tion between moral evil – the evil brought about by Hitler – and natural or physical evil – the Lisbon earthquake. The Christian re-sponse to moral evil is that of St. Augustine. God gave us free will: free will is a great gift for good, and it is better that we have it, even though we will then do evil, than that we do not have it and do nothing of our own accord (Augustine 1998). There are two issues here. First, does science as such – and Darwinism is part of science – make free will impossible? Second, is there something in Darwinism itself that makes free will impossible? The answer to the first question is that science and free will can go together, and indeed there are reasons to think that they must go together. David Hume (1739–40) is the authority here. If there are no laws governing human behavior, then we are not free – we are crazy, and do things without cause, without rhyme or reason. The compatibilist argues that the true dis-tinction is not between freedom and law, but between freedom and constraint. The person in chains is not free, nor is the person under hypnosis. It is true that they are subject to law, but so also is the free person not in chains or under hypnosis.

To take up the second question, does not Darwinian science specifically have something about it that puts us all under hypnosis – genetic hypnosis? Or, returning to a point that came up in the last chapter, as critics like Harvard biologist Richard Lewontin (1991) have put it, does not Darwinism deny freedom by making us "genetically determined"? We have no freedom, good or ill, because our genes make us act. Hitler is not to blame. He just had a lousy genotype (set of genes), and it is natural selection that put that in place. Blame the process not us. Likewise, of course, Mother Teresa is not to be praised. She drew a good genotype. However, this argument stands no more in the case of religion than it did in the case of philosophy. Some things are surely genetically determined. Ants, for instance. They are preprogrammed by the genes, as produced by selection. But we humans are not ants, we are not genetically determined in this way. Our evolution has been such as to give us the power to make decisions when faced with choices, and to revise and rework when things go wrong. Ants are like cheap rockets: many are produced and they cannot change course when once fired. We humans, by contrast, are like expensive rockets: just a few are produced but we can change course even in mid-flight, if the target changes direction or speed or whatever. The expensive rocket has flexibility – a dimension of freedom – not possessed by the cheap rocket. Both kinds of rockets are covered by laws, and so are ants, wasps, and humans. We have freedom over and above genetic determinism, and this freedom was put in place by – not despite – natural selection. Hence, the argument from evil against free will fails – at least, it fails if you are making the case based on Darwinism.

What about natural or physical evil? The traditional saving argument is one that is usually associated with the great German philosopher Leibniz. He pointed out that being all-powerful has never implied the ability to do the impossible. God cannot make $2 + 2 = 5$. No more can God, having decided to create through law (and there may be good theological reasons for this), make physical evil disappear. Physical evil simply comes as part of a package deal.

> For example, what would it entail to alter the natural laws regarding diges-
> tion, so that arsenic or other poisons would not negatively affect my con-
> stitution? Would not either arsenic or my own physiological composition
> or both have to be altered such that they would, in effect, be different from
> the present objects which we now call arsenic or human digestive organs?
> (Reichenbach 1976, 185)

Paradoxically, Dawkins himself rather aids this line of argument. He has long maintained that the only way in which complex adaptation could be produced by law is through natural selection. He argues that alternative mechanisms (notably Lamarckism) which produce adaptation are false, and alternative mechanisms (notably evolution by jumps, or saltationism) which do not produce adaptation are inadequate. "If a life-form displays adaptive complexity, it must possess an evolutionary mechanism capable of generating adaptive complexity. However diverse evolutionary mechanisms may be, if there is no other generalization that can be made about life all around the Universe, I am betting that it will always be recognizable as Darwinian life" (Dawkins 1983, 423). In short, if God was to create through law, then it had to be through Darwinian law. There was no other choice.

Conclusion

A critic of the God hypothesis might still argue that the problem of evil persists. Perhaps God should not have created the world, however good the ends, knowing the pain and suffering it would entail along the way. But this is another issue and not one raised by Darwinism. So let us draw now to a conclusion. Darwinism has certainly left its mark on traditional Christian thinking. Whether it has made any kind of Christian commitment impossible or unreasonable is a very different question. My own suspicion is that one can (if one wants) be both a Darwinian and a Christian. I cannot pretend that we have covered all of the bases. The reader might worry, with good reason, that in the last chapter, I rather left a gap between the moral dictates of Jesus and the supportable morality of the evolutionist. At least in part, my response to this is that there is a difference between the raw, unfiltered orders of Jesus of the Gospels and the demands of the Christian religion. Jesus tells us to leave our families for his sake, St. Paul tells us to look after them; Jesus tells us to turn the other cheek, St. Augustine works out the rules for a "just war." These later, clarified demands are much more in tune with the normative morality of the Darwinian. This is not a trivial or ad hoc response. It is clear now to biblical scholars that Jesus was talking in apocalyptic terms, expecting the end of time before his death. His followers had to work with the fact that his human nature misled him on this point, and that they had to pick up the pieces; this often involved modifying things (especially moral things) in directions that a Darwinian would expect (Ruse 2001).

Let us leave the discussion there for the time being. Some of these questions will come up again in the next chapter.

11

The Origins of Religion

As every enquiry, which regards religion, is of the utmost importance, there are two questions in particular, which challenge our attention, to wit, that concerning its foundation in reason, and that concerning its origin in human nature. Happily, the first question, which is the most important, admits of the most obvious, at least, the clearest, solution. The whole frame of nature bespeaks an intelligent author; and no rational enquirer can, after serious reflection, suspend his belief in a moment with regard to the primary principles of genuine Theism and Religion. But the other question, concerning the origin of religion in human nature, is exposed to some more difficulty.

These are the opening words of David Hume's essay *The Natural History of Religion*. It was written around 1749–51, and first appeared in print in 1757. It is odd, to say the least, to hear Hume sounding almost like an enthusiast for Intelligent Design theory, and one wonders somewhat about his sincerity on this issue. At the same time, he was also working on his *Dialogues Concerning Natural Religion*, in which – as we have seen – Hume was deeply critical of natural theology, even though, to the end, he was never an atheist and at most a skeptic. Our concern in this chapter, however, is with his second question – the origin of religion in human nature – and how it is to be answered in the light of Charles Darwin's theory of evolution.

Given that mine is a book about Darwin viewed from the perspective of philosophy, you might think that this is a topic to be skipped. Surely a question about origins is less a philosophical question and more an empirical question, to be answered by historians or sociologists or anthropologists, or perhaps even by evolutionists? For three reasons, it deserves our attention. First, whether or not it is truly a philosophical issue, it is certainly something that has captured the attention of philosophers – from David Hume in the Enlightenment to (most recently) Daniel Dennett. Second, not only did Darwin have things of interest to say on the topic, but today there is a huge amount of interest in the putative evolutionary origins of religion, so to omit all discussion would be unduly to truncate the overall picture of Darwinism today. Third, ultimately – as with just about everything – there are lurking philosophical questions worth considering. So let us plunge into the debate, starting with David Hume. Although he has many insights that cry out for an evolutionary interpretation, I do not think that Hume was an evolutionist. But he certainly set the terms of the debate.

David Hume

Hume offered a "natural" history of religion: that is, he explained religion entirely in natural terms – no miracles or hand of God or any such thing. He started by suggesting that polytheism was the original belief state of humankind and that it came from a tendency to see life in all things, including the inanimate. Primitive man was worried about food and security and all such things and this led him to interpret the world as though it were full of animate beings.

> There is a universal tendency among mankind to conceive all beings like themselves, and to transfer to every object, those qualities, with which they are familiarly acquainted, and of which they are intimately conscious. We find human faces in the moon, armies in the clouds; and by a natural propensity, if not corrected by experience and reflection, ascribe malice or good-will to everything, that hurts or pleases us.
>
> (Hume 1963, 40–1)

From here it was an easy enough thing to think that the whole world is filled with gods or deities of one sort or another. Hume added that women are often thought to have played a big role here, because it was they who were most given to doing this, and they infected men with their enthusiasms. Quoting the Greek philosopher, historian, and geographer Strabo, Hume says: "It is rare to meet with one that lives apart from the females, and yet is addicted to such practices" (p. 44).

266   *The Origins of Religion*

After he got going with the god idea, Hume then supposed that some divinities started to gain importance over all others, until we went all of the way to monotheism. Hume noted how in the Middle Ages the Virgin Mary was on the way to being promoted to the very top rank in people's minds – until the Protestant Reformation took the gloss off her glory. Not that Hume was much enthused by the end results: he inclined to think that the more a god gets in charge, the less tolerant are its supporters of rivals, and this leads to all kinds of tensions and frictions. Also, a more exalted god, elevated above the everyday nature of life, starts to take on attributes that reason suggests are absurd. Showing his good Calvinist training, Hume was particularly scornful of the Catholic doctrine of transubstantiation – the idea that we might be eating our god.

> Upon the whole, the greatest and most observable differences between a traditional, mythological religion, and a systematic, scholastic one are two: The former is often more reasonable, as consisting only of a multitude of stories, which, however groundless, imply no express absurdity and demonstrative contradiction; and sits also so easy and light on men's minds, that, though it may be as universally received, it happily makes no such deep impression on the affections and understanding.
>
> (p. 85)

One starts to get some sense of why David Hume made so many good believers very uncomfortable and why one critic (a fellow Scot) described him as God's greatest gift to the infidel.

Finally, before having the nerve to reaffirm the existence of a Creator and to go so far as to say that the "universal propensity to believe in invisible, intelligent power, if not an original instinct, may be considered as a kind of mark or stamp, which the divine workman has set on upon his work," Hume drew a firm line between belief in a divinity and moral behavior. If anything, there seems to be an inverse ratio, with greater intensity of belief leading to greater inclination to commit moral atrocities. People reason that being onside with God gives them license to do what they will.

> Hence the greatest crimes have been found, in many instances, compatible with a superstitious piety and devotion; Hence, it is justly regarded as unsafe to draw any inference in favor of a man's morals, from the fervor or strictness of his religious exercises, even though he himself

believe them sincere. Nay, it has been observed, that enormities of the blackest dye have been rather apt to produce superstitious terrors, and increase the religious passion.

(p. 94)

The Descent of Man

For all of my caginess about direct links between Hume and Darwin, in the matter of religion and its putative natural origins, the case for an immediate Humean influence is strong. Darwin turned to religion and its origins in *The Descent of Man*, and, while it is true that he did not footnote Hume (the *Descent*, being a more measured work, has footnotes, unlike the *Origin*, which was written at speed and intended to be an abstract of a larger whole), the naturalistic spirit is very much that of Hume. Years earlier, when Darwin was working toward natural selection and when he was thinking out his whole, overall position on evolution – including, especially, his position on humans (namely that we are as much subject to natural causes as any other organism) – he did read Hume's *The Natural History of Religion*. Although Darwin was obviously going to add an evolutionary perspective, a direct influence from Hume is more plausible here than at any other place in the Darwinian corpus.

What is striking about Darwin's discussion of religion is how brief it is, especially when compared to the detailed discussion that he gave of the origins and nature of morality. There are at least two reasons for this. First, by the 1870s, in the eyes of Victorians like Darwin, the battle for religion was over and God had lost. The problem now was to maintain morality in the face of nonbelief. The novelist George Eliot spoke for many when, discussing "God, immortality, duty," she declaimed: "how inconceivable . . . the first, how unbelievable the second, and yet how peremptory and absolute the third" (Myers 1881). Remember the similar passage from Kant that found its way into the *Descent* (Darwin 1871, 1: 70). Darwin gave much more space in the *Descent* to the evolution of morality. Second, and more importantly, Darwin was not a David Hume or (to speak of his contemporary) a Thomas Henry Huxley. His own belief may have gone, but by nature and class he was instinctively against religion-bashing. He had grown up with and in the church. His family was respectable and often religious – if not Anglican then Unitarian. One of his closest personal friends was the local vicar. Darwin was certainly not going to conceal his beliefs, but equally certainly he was not going to flaunt them.

Measured and brief or not, what did Darwin have to say?

> There is no evidence that man was aboriginally endowed with the ennobling belief in the existence of an Omnipotent God. On the contrary there is ample evidence, derived not from hasty travellers, but from men who have long resided with savages, that numerous races have existed, and still exist, who have no idea of one or more gods, and who have no words in their languages to express such an idea.
>
> (Darwin 1871, 1: 65)

To which sentiment, in Humean style, Darwin added immediately: "The question is of course wholly distinct from that higher one, whether there exists a Creator and Ruler of the universe; and this has been answered in the affirmative by some of the highest intellects that have ever existed." He also added that, if we are thinking in terms of vague spiritual sentiments, then obviously savages and other primitive folks come under the religion blanket. "If, however, we include under the term 'religion' the belief in unseen or spiritual agencies the case is wholly different; for this belief seems to be universal with the less civilised races" (1: 65).

How did this sentiment or belief about unseen forces arise? Here Darwin started to sound very Humean. It is all a question of seeing spirits in inanimate objects, feeling or pretending or mistakenly believing that they are truly alive.

> The tendency in savages to imagine that natural objects and agencies are animated by spiritual or living essences, is perhaps illustrated by a little fact which I once noticed: my dog, a full-grown and very sensible animal, was lying on the lawn during a hot and still day; but at a little distance a slight breeze occasionally moved an open parasol, which would have been wholly disregarded by the dog, had any one stood near it. As it was, every time that the parasol slightly moved, the dog growled fiercely and barked. He must, I think, have reasoned to himself in a rapid and unconscious manner, that movement without any apparent cause indicated the presence of some strange living agent, and that no stranger had a right to be on his territory.
>
> (1: 67)

From here, continuing in a Humean fashion, we are off and running toward a more sophisticated religious framework. Important is the notion of religious devotion – love of a god and so forth – and it is clearly animal in origin. Of course, you need to have developed a certainly level of intellectual power and sophistication, but "we see some distant approach to this state of mind in the deep love of a dog for his master, associated with complete submission, some fear, and perhaps other feelings" (1: 68). Like Hume, Darwin saw a move from primitive religion, through polytheism, and on to monotheism. And, completing the story still in the vein of the Scottish philosopher, Darwin saw religion as connected with vile superstitions and practices, which only the rise to reason could conquer and prevent.

> Many of these [superstitious practices] are terrible to think of – such as the sacrifice of human beings to a blood-loving god; the trial of innocent persons by the ordeal of poison or fire; witchcraft, &c. – yet it is well occasionally to reflect on these superstitions, for they shew us what an infinite debt of gratitude we owe to the improvement of our reason, to science, and to our accumulated knowledge.
>
> (1: 68–9)

And that, in the body of the text, was all Darwin was prepared to say on the subject, although at the end of the *Descent*, in summarizing he returned to the topic.

> The belief in God has often been advanced as not only the greatest, but the most complete of all the distinctions between man and the lower animals. It is however impossible, as we have seen, to maintain that this belief is innate or instinctive in man. On the other hand a belief in all-pervading spiritual agencies seems to be universal; and apparently follows from a considerable advance in man's reason, and from a still greater advance in his faculties of imagination, curiosity and wonder. I am aware that the assumed instinctive belief in God has been used by many persons as an argument for His existence. But this is a rash argument, as we should thus be compelled to believe in the existence of many cruel and malignant spirits, only a little more powerful than man; for the belief in them is far more general than in a beneficent Deity. The idea of a universal and beneficent Creator does not seem to arise in the mind of man, until he has been elevated by long-continued culture.
>
> (2: 394–5)

The Origins of Religion

Nor was Darwin prepared to say much about the immortality of the soul. You may believe in it, but do not pretend that the evidence from development is going to be much help. Savages do not have such beliefs, although, added Darwin comfortingly, one really should not read too much into this. Reflecting his status as an upper-middle-class Englishman, Darwin never thought that you should read too much into the beliefs or disbeliefs of primitive people.

Darwin's position, therefore, was that religion is a natural phenomenon – or rather a phenomenon that can be treated naturally – and he saw it as something that had evolved. Given the time that Darwin spent showing that morality is something deeply connected to natural selection, it is noteworthy that he did not at all attempt this in the case of religion. One presumes that since he thought it false – at least not proven in its essentials, and false in many details – he did not think that religion could be promoted by selection. It does not give us insights into reality, or (even if false) help us better to survive and reproduce. Nor did Darwin want to argue for religion as a byproduct of selection, or as something that might be promoted by selection but that lacked direct adaptive significance for survival.

For Darwin, religion seems to be almost accidental, and brought about by animal features or powers that are simply misdirected. When we see something moving, it normally makes sense to think that it is living. We make mistakes, and ultimately this leads into religion. About the only thing that can be said in its favor is that, in the case of civilized people, it does help to reinforce morality. "With the more civilised races, the conviction of the existence of an all-seeing Deity has had a potent influence on the advance of morality" (Darwin 1871, 2: 394). Here, perhaps, Darwin did differ from Hume, opting to be at one with his fellow Victorians, who, despite disbelief, were hymning the virtues of Bible reading as a guide to and reinforcement of morality. Huxley, the arch-agnostic, always advocated Bible study in state schools.

Darwinism and religion today

From the middle of the nineteenth century down to the present, much has been written on the subject of the putative natural origins of religion. Anthropologists and sociologists turned in a major way to religious practices and beliefs, in Western society and elsewhere in the world, in the present and in the past – the latter being sometimes discovered from evidence and sometimes created by lively imagination. Well known, perhaps "notorious" is a better term, are such comparative studies as J. G. Frazer's *The Golden Bough* (1890, 3rd edn. 1906–15), which sees religion as rooted in fertility cults, where the king must die and be reborn for the cycles of nature to move forward as they do; the implications of this for Christianity, a religion where the god dies and is resurrected, did not escape its author. Others also got into the business of supplying histories of religion and explanations of their attraction. Freud (discussing Moses and his supposed death at the hands of his followers) was typical:

> Religion is an attempt to master the sensory world in which we are situated by means of the wishful world which we have developed within us as a result of biological and psychological necessities. . . . If we attempt to assign the place of religion in the evolution of mankind, it appears not as a permanent acquisition but as a counterpart to the neurosis which individual civilized men have to go through in their passage from childhood to maturity.
>
> (Freud 1964, 168)

Freud was deeply interested in biology and thought that religion could be explained in biological terms. Deaths and sacrifices are echoes of actual events in the past, embedded in human nature by Lamarckian processes – the inheritance of acquired characteristics. Hence, even if Freud's claims about religion were correct – and many disputed such controversial ideas as that Moses was killed by his followers – the Lamarckian mechanism was wrong (Sulloway 1979). It was not until the 1970s that we see a genuine revival of attempts to explain religion in terms of modern evolutionary biology. Great credit must go to Edward O. Wilson, whose sociobiological synthesis made religion a central object of study; he discussed it in detail in his Pulitzer Prize-winning *On Human Nature* (1978).

Rather than structuring the discussion on the order of appearance of different ideas and hypotheses, let us be guided by biological categories – Darwinian categories, that is. Most importantly, we will expect a division between those who think that religion is brought about directly by natural selection, and those who think that it is a byproduct or spandrel. Then, among those who propose selection as the cause, there will be division between those who think that religion is of direct adaptive advantage to humans, and those who think that it might not be such a good thing to have, and who account for it as perhaps a product of sexual selection or as adaptive for someone or -thing other than humans (like parasites). There is also the possibility of a division between individual and group selectionists, and between those who think that religion is essentially biological and those who think that culture is significant, if not all-important.

Starting with those who think that religion is selection-produced and of value to humans, let us see what Edward O. Wilson himself says. Religion is apparently all a matter of group identity and sticking together.

> The highest forms of religious practice, when examined more closely, can be seen to confer biological advantage. Above all, they congeal identity. In the midst of the chaotic and potentially disorienting experiences each person undergoes daily, religion classifies him, provides him with unquestioned membership in a group claiming great powers, and by this means gives him a driving purpose in life compatible with his self interest.
>
> (Wilson 1978, 188)

Wilson does allow that there can be cultural selection between sects, but essentially we start with the biology and all else is on the surface.

> Because religious practices are remote from the genes during the development of individual human beings, they may vary widely during cultural development. It is even possible for groups, such as the Shakers, to adopt conventions that reduce genetic fitness for as long as one or a few generations. But over many generations, the underlying genes will pay for their permissiveness by declining in the population as a whole. Other genes governing mechanisms that resist decline of fitness produced by cultural evolution will prevail, and the deviant practices will disappear. Thus culture relentlessly tests the controlling genes, but the most it can do is replace one set of genes with another.
>
> (p. 178)

Wilson has always been ambivalent about the comparative significance of individual- and group-selective processes, or more precisely has always been more tolerant of, and open to, group levels of action than are many evolutionary biologists. On the religion issue, he is rather divided, thinking that it is brought on by a group process, but surely

The Origins of Religion   273

has individual benefits also. More robustly individualistic are the physical anthropologist Vernon Reynolds and the scholar of religion Ralph Tanner (1983). They accept hypotheses such as that circumcision of males – which is central to religious practices of Jews and others – originated as a measure to prevent disease. This is a practice that benefits individuals. Somewhat ingeniously, Reynolds and Tanner suggest that religions tend to divide into those that promote high reproductive rates – many Semitic religions – and those that do not – north European Calvinism for instance. As it happens, this echoes interests and concerns of Darwin in the *Descent*. Remember how the great evolutionist worried that the worthless Catholic Irish seemed to have lots of children, whereas the hardworking Presbyterian Scots had but few. This was a horrific reflection, which seemingly negated the upward, progressive nature of the evolutionary process, a picture so dear to the heart of Darwin and his fellow Victorians.

Remember also that Darwin consoled himself with the reflection that the Irish do not look after their kids whereas the Scots do, and so, on balance, the Scots if anything do better than the Irish. Reynolds and Tanner draw on r and K selection theory, fleshing out this intuition; they suggest that if conditions are highly variable then a good reproductive strategy is to have lots of offspring, albeit they receive little individual attention, whereas if conditions are stable then the better strategy is to have few children and give them much care. Variable conditions mean that with lots of offspring sometimes you might strike it rich, whereas with few offspring you might never do that well. And conversely. Remember that K selection works for stable conditions and r selection for unstable conditions. Humans and elephants are overall K-selected whereas herrings and mice are overall r-selected. Reynolds and Tanner refer to these two modes as r^c- and r^c+ selection, the superscript signifying that they think the selection has to work through culture, although presumably it filters down to the genes.

In the case of male circumcision, acknowledging that there are questions about the evidence, Reynolds and Tanner nevertheless write:

> Despite the confused state of the data, it is not unreasonable to put the question: If circumcision does reduce the risk of penile or cervical carcinoma, what effects would this have on reproductive success? The answer is that such success should be increased (all other things being equal) in

The Origins of Religion

families or groups practising circumcision. Circumcision would thus be a pro-reproductive practice and should be favoured in situations in which r^c+ selection was operative. We know that it is a characteristic of long-standing in Judaism and Islam. In the case of Judaism it represents part of Abraham's Covenant with God, the covenant in which God called him to leave Ur and to found a new nation; also in the Covenant was the promise from God that his 'seed' would inherit the land. A charter for r^c+ selection indeed! In the case of Islam, circumcision appears to have been simply continued without question from a prior Arabic tradition. The practice is not mentioned at all in the Koran and was adopted without question by Muhammad; it is regarded as an essential of the faith.

(Reynolds and Tanner 1983, 240)

Groups and memes as units of selection

Edward O. Wilson and Reynolds and Tanner are fairly conventional Darwinians. However, showing just how different people's thinking can be and yet still (in the eyes of advocates) be under the banner of Darwinism, we have the biologist David S. Wilson and the philosopher Daniel Dennett. As one might expect from earlier discussion, David S. Wilson (2002) is openly committed to a group-selective analysis of religion, wanting to regard societies as akin to organisms and as strengthened by a sincere commitment to a religious doctrine. He ties this thesis strongly to morality, which he speaks of as having "both a genetically evolved component and an open-ended cultural component" (p. 119). Wilson analyses the society that Jean Calvin founded in Geneva in the sixteenth century, listing the rules that governed this group: "Obey parents"; "Obey magistrates"; "Obey pastors"; and on down the list to "No lewdness and sex only in marriage"; "No theft, either by violence or cunning"; and so forth. Of this he writes:

To summarize, the God–people relationship can be interpreted as a belief system that is designed to motivate the behaviors [examples of which are listed just above]. Those who regard religious belief as senseless superstition may need to revise their own beliefs. Those who regard supernatural agents as imaginary providers of imaginary services may have underestimated the functionality of the God–person relationship in generating real services that can be achieved only by communal effort. Those who already think about religion in functional terms may be on the right track, but they may have underestimated the sophistication of the "motivational physiology" that goes far beyond the use of kinship terms and fear of hell. Indeed, it is hard for me to imagine a belief system better designed to motivate group-adaptive behavior for those who accept it as true. When it comes to turning a group into a societal organism, scarcely a word of Calvin's catechism is out of place.

(Wilson 2002, 105)

Although he (like Edward O. Wilson and Reynolds and Tanner) thinks that culture is crucial, one suspects that ultimately David S. Wilson (and they) see real change as genetic. Coming from a very different direction, Daniel Dennett (2006) agrees entirely that religion is promoted by selection, but he is not at all convinced that this selection is necessarily for the benefit of humans, nor is it essentially (or, truly, in any way) genetic. Dennett has adopted a theory of Richard Dawkins (1976) that posits the existence of "memes," units of culture akin to genes, which compete for people's allegiances. Rival memes, as it were, invade people's minds and those that win are those that are selected to continue. Winning is not random, but a function of the features – the adaptations – that the memes have or promote. Successful advertising obviously is a paradigmatic example of memes at work: you buy and smoke Marlborough cigarettes because this makes you feel like a real man, even though, in fact, you are acting in ways highly detrimental to your health and well-being.

For Dennett, religion is a meme parasite, which has features that make it attractive, even if it is not necessarily that good for the possessor. So that no one should miss this point, he begins his book *Breaking the Spell: Religion as a Natural Phenomenon*, by introducing the reader to the lancet fluke (*Dicrocelium dendriticum*): this parasite corrupts the brain of an ant, causing it to strive to climb blades of grass, so that it gets eaten by a sheep or cow, and thus the fluke can complete its life cycle before its offspring are excreted and take up again with ants.

Does anything like this ever happen with human beings? Yes indeed. We often find human beings setting aside their personal interests, their health, their chances to have children, and devoting their entire lives to furthering the interests of an *idea* that has lodged in their brains. The Arabic word *islam* means "submission," and every good Muslim bears witness, prays five times a day, gives alms, fasts during Ramadan, and tries to make

276   *The Origins of Religion*

the pilgrimage, or *hajj*, to Mecca, all on behalf of the idea of Allah, and Muhammad, the messenger of Allah. Christians and Jews do likewise, of course, devoting their lives to spreading the Word, making huge sacrifices, suffering bravely, risking their lives for an idea. So do Hindus and Buddhists.

(Dennett 2006, 4)

To be fair, Dennett adds that secular humanists are often not much better in this regard.

As we might expect, Dennett is disdainful of the suggestion that religion has much to do with morality.

I have uncovered no evidence to support the claim that people, religious or not, who don't believe in reward in heaven and/or punishment in hell are more likely to kill, rape, rob, or break their promises than people who do. The prison population in the United States shows Catholics, Protestants, Jews, Muslims, and others – including those with no religious affiliation – represented about as they are in the general population.

(p. 279)

In other words, although they would both claim to be Darwinians, when it comes to religion and its virtues, David S. Wilson and Daniel Dennett are about as far apart as they could be.

Religion as byproduct

Turning now to the other side of the equation, what of those who think religion falls more into the spandrel category? Gould (2002) certainly thinks somewhat along these lines: the whole of human culture comes under this category for him. But most would not be so sweeping. Apart from anything else, religion with its costs – devotion to others, celibacy, ritual physical disfigurement, sacrifice, and so forth – simply does not seem to be the sort of thing that would happen as a byproduct. It is just too costly. More likely, in the opinion of some, is the idea that religion piggybacks, as it were, into existence and power by attaching itself to other things – real, powerful adaptations – and manages to exist because it cannot be stopped or because ultimately its costs are simply not that great. Student of culture Pascal Boyer inclines to the first option. For him, religion simply subverts or borrows features that our biology has put in place for good adaptive reasons, and for whatever reason it cannot be eradicated.

> The building of religious concepts requires mental systems and capacities that are there anyway, religious concepts or not. Religious morality uses moral intuitions, religious notions of supernatural agents recruit our intuitions about agency in general, and so on. This is why I said that religious concepts are parasitic upon other mental capacities. Our capacities to play music, paint pictures or even make sense of printed ink-patterns on a page are also parasitic in this sense. This means that we can explain how people play music, paint pictures and learn to read by examining how mental capacities are recruited by these activities. The same goes for religion. Because the concepts require all sorts of specific human capacities (an intuitive psychology, a tendency to attend to some counterintuitive concepts, as well as various social mind adaptations), we can explain religion by describing how these various mind capacities get recruited, how they contribute to the features of religion we find in so many different cultures. We do not need to assume that there is a special way of functioning that occurs only when processing religious thoughts.
>
> (Boyer 2002, 311)

But what is it that allows religion to get its hold in the first place? Anthropologist Scott Atran inclines to the second option, that religion grabs something adaptively useful and exploits it. For him, the big question facing organisms like humans is other living beings – above all, other living beings as threats. In an argument reminiscent of Darwin and his dog – in fact, reminiscent of Hume's speculations – Atran suggests that what we have is a somewhat overeager projection of the living onto the inanimate. It used to be thought that the baroque nasal appendages of the titanotheres were a case of sensible evolution having taken a step too far. Perhaps the same is true of religion. Cuckoos exploit the innate mechanisms that their host birds have for raising their young. Religion does much the same for humans.

> Supernatural agent concepts critically involve minimal triggering of evolved agency-detection schema, a part of folkpsychology. Agency is a complex sort of "innate releasing mechanism." Natural selection designs the agency-detection system to deal rapidly and economically with stimulus situations involving people and animals as predators, protectors, and prey. This resulted in the system's being trip-wired to respond to fragmentary

The Origins of Religion

information under conditions of uncertainty, inciting perception of figures in the clouds, voices in the wind, lurking movements in the leaves, and emotions among interacting dots on a computer screen. This hair-triggering of the agency-detection mechanism readily lends itself to supernatural interpretation of uncertain or anxiety-provoking events.

People interactively manipulate this universal cognitive susceptibility so as to scare or soothe themselves and others for varied ends. They do so consciously or unconsciously and in causally complex and distributed ways, in pursuit of war or love, to thwart calamity or renew serendipity, or to otherwise control or incite imagination. The result provides a united and ordered sense for cosmic, cultural, and personal existence.

(Atran 2004, 78)

Richard Dawkins fall somewhere in this category. Predictably, he has little time for the supposition that morality and religion have much connection, quoting with relish the Old Testament stories of the behavior of the patriarchs. Of Abraham and Isaac he writes: "By the standards of modern morality, this disgraceful story is an example simultaneously of child abuse, bullying in two asymmetrical power relationships, and the first recorded use of the Nuremberg defence, 'I was only obeying orders'" (Dawkins 2006, 242). Since, for him, religion has only a negative value, he too inclines to the religion-byproduct thesis. Although he thinks the general principle more important than the details, he wonders if the adaptive advantages of obeying informed authority – "stay away from the cliff" – have been subverted by the religion memes – "do this, that, and the other, or you will go to hell."

Serious science?

What can we say about these various ideas and hypotheses? One thing for certain. They can't all be true! For every action there is an equal and opposite reaction. For every idea about the evolution of religion, there is an idea that takes exactly the opposite tack. You think, with both Wilsons, that religion evolved and helps humans? Then what of Dennett, who likens it to the fluke parasite – hardly the ultimate model of warmth and friendliness. Or consider Atran who thinks that religion is a case of adaptation going over the top. As he says explicitly: "supernatural agency is an evolutionary by-product, trip-wired by predator-protector-prey detection schema." You think, with Reynolds and Tanner, that religion evolved to help the individual? Then read David S. Wilson, who is convinced that it is groups all of the way. You think, with Edward O. Wilson, that religion is a function of biological evolution? Then look at Dennett who thinks that it is culture – simple, albeit not very pure, culture. You argue with David S. Wilson that religion is the ultimate support of morality? Then take up Dennett (not to mention David Hume), who thinks that religion does absolutely nothing for morality, or, if it does, its effects are almost invariably negative. Amusingly, Dennett quotes the Nobel physicist Steven Weinberg: "Good people will do good things, and bad people will do bad things. But for good people to do bad things – that takes religion" (Dennett 2006, 279). Although Dawkins thinks Jesus is a step up from the figures of the Old Testament, he criticizes him for "his somewhat dodgy family values," and the death on the cross is "a new sado-masochism whose viciousness even the Old Testament barely exceeds" (Dawkins 2006, 250–1).

I do not want to argue that it is necessarily impossible or worthless to invoke Darwinian evolutionary theory in the explanation of the origin or nature of religion. I do rather argue that we have got to do a lot better than we have done thus far. Part of the issue, obviously, is with the general program of human sociobiology, or evolutionary psychology, or whatever you want to call it. As and when this improves overall, so biological explanations of religion as a part of human culture will get dragged along too. Part of the issue equally obviously lies in teasing out which causal explanation is generally better in these circumstances. Leave aside the individual-selection/group-selection discussion as something that has been considered in depth already, and take what Dennett relies on, the theory of memes, or memetics, or whatever it is called. It really is crude to the point of non-being, certainly to the point of non-helpful. What is a meme? It is a chunk of culture analogous to a gene. As it happens, genes are hard enough to define, but we do have some idea of them as the smallest functioning length of DNA. But what is the smallest functioning length of culture? Is Catholicism a meme? Is the authority of the Pope a meme? Is transubstantiation a meme? Why the authority of the Pope, for example, rather than each and every one of the dogmas that he endorses? And what kind of theory do you have as memes clash and come together and sometimes fuse and sometimes break apart? How is Mormonism a meme as compared to evangelical Christianity? Does Mormonism somehow include a lot of the evangelical Christianity meme, or are they separate memes? And so on and so forth.

The point is not that Dennett is necessarily wrong in arguing that ideas sometimes have lives of their own, or that religions can be dreadful things that take over people's minds to their own detriment – if we do not think this way about Catholicism then most of us do about cults like Scientology – but that memetics is not very helpful in understanding what is going on. One is really just taking regular language and putting it in fancy terms. No new insights. No new predictions. No astounding claims that turn out to be true. More importantly, one is not really using Darwinian evolutionary theory to do any work. One could be a Lamarckian and make most of the claims that Dennett wants to make. In fact, given the way that memes can be transmitted from one individual to another, one would probably be more comfortable being a Lamarckian if one wanted to make most of the claims that Dennett wants to make.

And part of the issue here – part of the problem with Darwinian approaches to religion – is simply that so much of the scholarship is so crude. Let me praise David S. Wilson for wanting to use real examples to articulate and flesh out his thinking. But his discussion of Calvinism really will not do. If Calvinism was such a terrific booster of societies and helped them work so well, why did it so frequently fail to convince? Take the English (MacCulloch 2004). Henry VIII broke from the Catholic Church because he wanted to take a new wife and the Pope forbad it. His son, Edward VI was ultra-Protestant. But when Edward died as a teenager his elder sister Mary came to the throne and, as an ardent Catholic, persecuted Protestants, many of whom fled to the continent. By that time, the middle of the sixteenth century, the German Lutheran areas were torn by war and strife, and so these exiles headed for safer, quieter, Reformed (Calvinist) areas. When Mary died and her younger Protestant sister, Elizabeth, came to the throne, the Calvinists all flooded back. But generally, the English were not that keen on what Calvinism had to offer. They did not want the repressive morality and lifestyle of those who later came to be known as Puritans. So the Elizabethan compromise was worked out in the shape of an Anglican Church – to this day a funny brico-lage of Catholic style and Protestant theology. But it was certainly not unsuccessful: the English saw off the Spanish and their armada. It is true that in the seventeenth century the Roundheads, the Puritans, won the Civil War and lopped off the head of Charles I, but within twelve years the Royalists, the more central Anglicans, were restored and the Puritans were out again.

The point I am making is that you cannot just isolate one bit of history, one place in time and space, and think you have the basis for a universal theory. You have got to spread your grasp much more broadly and confront the difficult cases, the examples that contradict your argument. This comment applies particularly to Americans – Edward O. Wilson and Daniel Dennett come at once to mind – who start with assumptions about the universal appeal and force of religion. By any measure, given its anti-Enlightenment obsession with religion, America is a very peculiar country, at least compared to the rest of the First World. It is very dangerous to argue about the need that humans have of religion if, in fact, a lot of humans really do not seem to need that much religion at all. England is a case in point. Most young couples want a church wedding. After all, they want things done properly. But do they spend their subsequent married life obsessed with the death of Jesus Christ on the cross as a payment for their sins? Such a claim would be laughable. They probably have some vague notion that the life hereafter will be a kind of extended weekend, with lots of telly and visits to the pub; but generally, to the lives of the average person in England, religion is about as relevant as the royal family. Significantly, the Queen and her children are expected to observe the ritual practices as a proxy for the rest of the population. I am not saying that all writers – not even all American writers – on the biology of religion are equally provincial in thinking that their home society is the norm, but the issue is inadequately addressed.

The point is made. The biology of morality has and is making significant strides. One cannot really say the same about the biology of religion.

God?

Quality science or not, the idea that there might be some connection between biology and religious belief makes some people distinctly edgy. When asked for his opinion on the "God gene," a stretch of DNA that one geneticist has recently associated with religious belief, the

The Origins of Religion

Reverend Sir John Polkinghorne, FRS, sometime physicist, Anglican clergyman, former President of Queen's College, Cambridge, and winner of the Templeton Prize for Progress in Religion, replied with some asperity: "The idea of a god gene goes against all my personal theological convictions. You can't cut faith down to the lowest common denominator of genetic survival. It shows the poverty of reductionist thinking" (quoted in the *Daily Telegraph*, November 14, 2004). I warned you earlier that for some people "reductionism" is associated with the Dark Arts – a bit like doing rude things with virgins in churches at midnight.

Suppose, however, for the sake of argument, that there is something to the naturalistic Darwinian approach to religion, its history and its nature. What philosophical implications does this have? What does this tell us about God, his nature and his existence? Are Polkinghorne's worries (and fears) justified? You might flip the argument entirely on its head, showing on non-Darwinian grounds that God does not exist and then setting forth on a naturalistic journey to explain why nevertheless so many people persist in believing that he is real. This is the tack taken by Dawkins, as well as his close supporter Daniel Dennett. The latter trots through the various arguments for the existence of God, and follows through with the standard objections. (No fault in that: the standard objections are pretty good.) Then he adds a twist of his own, or more accurately adds a twist of David Hume's – namely, that the very idea of God in the Christian sense is incoherent. Talking of science and its methods, he says:

It is only because I am confident that the experts really do understand the formulas that I can honestly and unabashedly cede the responsibility of pinning down the propositions (and hence the responsibility of understanding them) to them. In religion, however, the experts are not exaggerating for effect when they say they don't understand what they are talking about. The fundamental incomprehensibility of God is insisted upon as a central tenet of faith, and propositions in question are themselves declared to be systematically elusive to everybody. Although we can go along with the experts when they advise us which sentences to say we believe, they also insist that they themselves cannot use their expertise to prove – even to one another – that they know what they are talking about. These matters are mysterious to everybody, experts and laypeople alike.

(Dennett 2006, 220)

All of this really makes nonsense of the God hypothesis. And hence there is need of a Dennett-like naturalistic account of the origins of religion to show how it has taken a grip on human minds and culture. It is hardly surprising that for Dennett religion is a parasite like the lancet fluke. For Dawkins, it is child abuse.

I will return in a moment to Dennett, but we must first ask the basic question about God and his existence – not whether God exists but whether a Darwinian account of origins shows that God exists or not. Whether or not he was writing sincerely, David Hume separated out a naturalistic account of religion's origins from its truth status, and this seems also to be Darwin's approach. Edward O. Wilson goes entirely the other way. Darwinism gives a naturalistic account of religion and that is end to religion as a reliable authority on what there really is. As it happens, Wilson thinks that the human psyche demands religion, and hence at this point he feels free to substitute his own kind of evolutionary humanism. But this is because Darwinism has already done its corrosive work.

> But make no mistake about the power of scientific materialism. It presents the human mind with an alternative mythology that until now has always, point for point in zones of conflict, defeated traditional religion. Its narrative form is the epic: the evolution of the universe from the big bang of fifteen years ago through the origin of the elements and celestial bodies to the beginnings of life on earth. The evolutionary epic is mythology in the sense that the laws it adduces here and now are believed but can never be definitively proved to form a cause-and-effect continuum from physics to the social sciences, from this world to all other worlds in the visible universe, and backward through time to the beginning of the universe. Every part of existence is considered to be obedient to physical laws requiring no external control. The scientist's devotion to parsimony in explanation excludes the divine spirit and other extraneous agents. Most importantly, we have come to the crucial stage in the history of biology when religion itself is subject to the explanations of the natural sciences. As I have tried to show, sociobiology can account for the very origin of mythology by the principle of natural selection acting on the genetically evolving material structure of the human brain.
>
> If this interpretation is correct, the final decisive edge enjoyed by scientific naturalism will come from its capacity to explain traditional religion, its chief competition, as a wholly material phenomenon. Theology is not likely to survive as an independent intellectual discipline.
>
> (Wilson 1978, 192)

The Origins of Religion

As is inevitably the case when people start philosophical arm-wrestling with the crafty Scottish genius, they lose. Wilson is wrong and Hume is right, whatever the latter really thought about the God question. The fact that you can give a naturalistic explanation of religion does not at once, or at all, imply that religion is false. I can give a naturalistic explanation of my belief that the truck is bearing down on me, but it does not follow that the truck is not bearing down on me. It is true that if all you have is a naturalistic explanation, then (Dennett-and-Dawkins-like) you are probably not eager to embrace religion. If can you show that religion is indeed a parasite on the mind, why take it any more seriously than the hucksters' email claims that their potions will increase your penis length? But for the traditional religious person – at least, for the traditional Christian religious person – religion has another source of epistemic power that email spam does not have. Faith. This is something that comes first, second, and third. This being so, then far from a naturalistic account being threatening, many expect such a naturalistic explanation of origins. God had to impart the information to humankind in some way, and why not through evolution? Nor would it be a counter-argument that the explanation might make the arrival of religion rather less than edifying – perhaps that it is not a direct function of selection but a spandrel or something. The point is that the job is done. As the Victorians were always pointing out, if you can accept that God designed human reproduction to work in the way that it does, who is going to cavil at a little (or a lot of) evolution?

What about the issue that came up in the metaethical discussion? The argument against the objective nature of morality was that, in theory, we could have another morality (the Cold War system of morality) contradicting the one we do have. But in religion we already have all of this and more! The Christians believe one thing, the Jews another, and the Muslims a third. Now, you might think that this is a pretty good argument against all of them. How can the Christian God be so loving and insist that we acknowledge and worship him, while condemning to eternal damnation all of those Asians who grew up in ignorance? But even if you do accept the argument against God based on comparative religion – I myself find it pretty convincing – note that this has nothing whatsoever to do with evolution. It was an argument that moved the deists at the end of the seventeenth century. Moreover, evolution or not, the believer can continue to believe in the face of religious diversity: the Christians (or whoever) got it right and the others did not, and that is the end of matters. The Calvinist God is notorious for condemning some humans, even though they are what they are by chance. What is a gene-given homosexual orientation to God if one is a practicing sodomite?

Of course, an alternative move in the face of conflicting claims is to try to extract some common theme from all religions – a God who appears to different people in different guises, all equally valid: Jesus for Christians, Buddha for Buddhists, and so forth. It is here that someone like Dennett would strike. By now, the believer has got so vague and inconsistent that really the object of his or her faith is not worth taking seriously. In an earlier book, *Darwin's Dangerous Idea* (1995), Dennett is particularly eloquent on the subject of rationality, pointing out that in a court of law when we are accused of a crime we are all in favor of rationality. Hence, when it comes to religion and God we should be equally in favor. Even without other religions, the Christian God is hopelessly vague and mysterious. Now, with the other religions, the whole God notion collapses in contradiction and absurdity.

To which again one can reply that this may be so, but it is hardly a consequence of evolution, of Darwinism. The argument holds whatever you think about origins. And perhaps here is a point when one can bring evolution, Darwinism specifically, to the aid of the Christian. Remember the point of Richard Dawkins about how we humans have adaptations for getting out of the jungles and onto the plains, not for peering into the mysteries of the universe. The limits to our understanding certainly do not prove that Christianity or any other religion is right, but perhaps they make a place for religion. Perhaps they show that the Christian is not entirely irrational in thinking that there might be more to existence than as we conceive it. "For now we see through a glass, darkly; but then face to face: now I know in part; but then shall I know even as also I am known." Modern sophisticated religions do end in mystery, but perhaps this is less an evasion than a reflection of our limitations and our humility before them. St. Paul, writing to the Corinthians, beat Dawkins to the punch by 2,000 years. And, with this reflection, let us leave naturalistic accounts of religion.

12

The Darwinian Revolution

Let me start this chapter on a personal note. Many years ago, in 1979, I published a book called *The Darwinian Revolution*. It was about Charles Darwin and the *Origin of Species* and the coming of evolution, and about the relative contributions of Darwin and others to its discovery. In my book I built on the labors and interpretations of scholars who had been working on the topic for the two decades since the centenary of the *Origin of Species*, a date (1959) that marked both the coming of age of the history of science as a scholarly discipline and the first systematic use of the massive Darwin archives in Cambridge to throw light on Darwin and the movement associated with his name. As is the nature of these things, my book was praised by some and criticized by others. But no one doubted that I had a topic. There was a Darwinian Revolution and my book was about it.

That was all back then. Today, a lot of people wonder if there was a Darwinian Revolution at all, and if so what one can and should say about it. Drawing together many of the threads of this present book, I will now turn to this issue, and ask three questions. First: was there a Darwinian *Revolution* – meaning, something that counts as a revolution – or is this kind of talk unhelpful? Second: was there a *Darwinian* Revolution – meaning, what role did Darwin himself play in any such revolution? Third: was there a *Darwinian Revolution* – meaning, what would be the nature, especially the philosophical nature, of any events that occurred in the nineteenth (and other) centuries that are linked with the name of Charles Darwin?

Was there a Darwinian Revolution?

Let us start right at the beginning. A number of scholars today say that all talk of "revolutions" is mistaken and misleading. In adopting this metaphor one is applying categories of today to the past, and forcing the past into structures which are not really appropriate or informative. In particular, it is a mistake to talk of scientific revolutions, and we should drop all such talk. There was really no Scientific Revolution in the sixteenth and seventeenth centuries and there was really no Darwinian Revolution in the nineteenth century. Of course, things happened, but they were not revolutionary. This does not mean to say that they were not significant. Obviously there were significant events in both physics and biology, but highlighting certain episodes as "revolutionary" has implications both about importance and about the special nature of the events that is really not warranted. Using the term means that you set out to find what may not be there, and, with scholars everywhere, a little problem like non-existence is not going to stand in the way of success. Even if one allows the term "revolution" for political events – the American Revolution, the French Revolution (even here a bit misleading because they were two very different events) – it is not helpful to transfer the term to science.

The leading debunker of the Darwinian *Revolution* interpretation – what he refers to amusingly as the "evo-revo" school – is the eminent historian Jonathan Hodge (2005), who states flatly that "historians of science should abandon any notion of a Darwinian revolution." He thinks it misleading because, in some sense, it focuses us on the Darwinian period, when we should be looking at the whole history of evolutionary theory. It forces us to think that this was when the really significant action occurred, when we should be realizing that it was one episode among many. Hodge believes that the very notion of a Darwinian Revolution was an invention of Darwin's supporters after the *Origin* was published. They wanted to inflate Darwin's (and, by implication, their own) importance, and what better way than by labeling what was happening as revolutionary? Hence, talk of a Darwinian Revolution was a function of propaganda needs rather than serious conceptual analysis. Drop it!

Let us sort out the gold from the dross here. If the point is that by focusing on the Darwinian Revolution we ignore or trivialize the rest of the history of biology, of evolutionary biology in particular,

The Darwinian Revolution

then (supposing it is true) this is a legitimate complaint. We should certainly not force that whole history into one short time period, or at least we should not do so simply because of our metaphor without having thought the matter through. And, certainly, one has to agree that there is some truth in what the critics say. I myself went straight to Darwin because of the "obviousness" of the Darwinian Revolution. I was (and am) primarily a philosopher. I focused on Darwin because people (leading philosophers of the day) told us that the key episodes in the history of science were revolutionary, and Darwin was the big figure in biology. Why did I not go first to Aristotle, say, who was a simply brilliant biologist? It was not just nationalistic chauvinism (I am English-born) but because that was where I "knew" the action lay.

Having said that, however, it is simply not true that today the rest of the history of evolutionary biology is ignored. There is an ever growing literature on Aristotle, for instance (Lennox 2001) – even by me (Ruse 2003)! The years before Darwin have been much explored and the post-Darwinian era even more so. In recent years, historians have been giving the twentieth century detailed treatment (just check out the trade magazine, the *Journal of the History of Biology*). The science of the last few generations is undergoing very careful investigation (though one wonders, given the ephemeral nature of much of today's electronic communication, whether it will be possible in the future to do the same for today's science).

Another version of the critics' complaint is that one should not use a term like "revolution" because it means so many different things – or applies to so many different events – that it becomes trivial and misleading. But this is surely a matter of style and method rather than substance or essence. All would agree that we should not assume that one revolution is going to be exactly like another. Especially in science, we should not assume that all revolutions are alike. But the word "revolution" does have a standard meaning, and it is useful. It means a dramatic change from one state to another. The American Revolution was certainly revolutionary in this sense: before, the country was ruled by the British; after, it was not. This made all of the difference. Similarly in other cases. The Information Revolution makes a lot of sense. It was not long ago that one had to book ahead to make a telephone call on Christmas day to England from Canada. Now, you can always pick up the phone and dial straight through. Many of us have had the experience of doing radio or television interviews in one country, with other participants in other countries, and the main studio and anchor somewhere else again. Anybody over forty sitting at their laptop today, grabbing information from the internet, needs no more proof that something pretty revolutionary has occurred than to remember what it was like to do research when they were young.

Of course, this in itself does not imply that the Darwinian Revolution was revolutionary, but by any measure it surely was. If you are not convinced by this stage of this book, you never will be! At the beginning of the nineteenth century, by and large, people did not believe in evolution. At the end of the nineteenth century, by and large, people did believe in evolution. More than this, they accepted that it applies to our own species, *Homo sapiens*. This was a terrific shift. Certainly, it was not purely a scientific revolution. Perhaps it was not even primarily a scientific revolution, being more to do with religion – Does God still exist and what does he care about us? – or culture, or whatever. But it was a revolution and, moreover – whether it was overall the most important factor or not – science was a very important factor. I myself would go further and say that it was the prime causal factor, for without the scientists I do not see how you could have had a shift to what is (after all) a scientific claim: organisms, including humans, evolved. And – without getting into discussion about whether all revolutions have to be major – this was the kind of massive change of mind that we associate with revolutions. For those of us who take the implications seriously, it matters more than just about anything else. Certainly more than anything else we can find out about ourselves.

Whiggism

So the critics go too far here. But I suspect that underlying their complaints is a third, deeper, objection. The critics are professional historians. The one thing that is drummed into professional historians today is that you must not judge the past by the present, above all you must not assume that the present is better and the past is worse. This is the dreadful sin of "Whiggism," a term taken from histories written back in the nineteenth century and earlier, supposedly proving that everything led up to the glories of the Whig party. By highlighting the revolutionary nature of Darwinism, you are forcing the nineteenth century into today's categories, and, moreover, portraying what happened as a move from dark to light. Dreadful!

Academics are a bit like sheep. One person says something, and before long the whole flock is bleating in unison. What is wrong with judging the past in terms of the present? We need some proof, some arguments, before we follow the jingling bell of the wether at our head. Let us grant that one should not simply mine the past for support for the present – the kind of thing that one often sees at the beginning of science textbooks: Darwin, Mendel, double helix. If the present needs support and the past can provide it, then let the links be spelt out. Let us also grant that doing history does require an ability to go back in time and to think things through in their terms rather than ours. Take the question of adaptation. Leave to one side whether or not you think it is overblown in modern evolutionary theory. Try first to understand it in terms of Darwin and Huxley. Why was it that Darwin thought it a key issue and Huxley did not? Unless you understand something of education back then, of how a university man like Darwin would think about design, and how a non-university man like Huxley, who was educated at a medical school (in those days a trade school outside the university system), would think about design, you are going to get nowhere in trying to grasp the differences. You may have the same views on God as does Richard Dawkins, but unless you put those to the back of your mind when you think about Darwin discovering natural selection, you are simply going to miss what was happening.

Having said this, however, the very act of interpretation is not bad history. Indeed, it is essential for history. Without interpretation there is just one fact after another – chronology. Darwin did this. Huxley did that. The world said: "A plague on both of them." And so forth. We want to know why the world said: "A plague on both of them." Or, more interestingly and accurately, why the world did not say: "A plague on both of them." Why was it that the anonymous author of *Vestiges of the Natural History of Creation* was reviled, whereas Huxley became a force in Victorian society, ending as a Privy Councilor? The facts are needed, but then interpretation is essential. Yet, you might respond, there are interpretations and interpretations. The problem with revolution talk is that it automatically implies progress, that things are getting better. The American Revolution is rightly so called because, after the British had been thrown out, there was a better society than before. And in the case of science, you are implying that later science is better than earlier science. Perhaps it is. Perhaps it is not. But you should not assume that it is, as you do at once by using revolution talk. As a historian it is not really your job to make these evaluations. Say how it went, not that it was better. A history of the Reformation would be devalued if the author assumed that the Protestant theology was better than the Catholic theology.

In response, for a start I do not see that necessarily one implies that things get better because of revolutions. The Russian Revolution of 1917 seems to me to be a clear counter-example to this argument. Many (I am not one, except when a student emails me when I am on vacation in Paris) think that the Information Revolution has been altogether too much of a good thing. I am even prepared to argue about the American Revolution – I would love to see little American children going to school in blazers, with caps and ties – although perhaps that is a topic better left to a book on its own. But what about science? If you genuinely do not think that science advances, then personally I believe that you have been reading too much philosophy. Of course it does! The earth is not 6,000 years old and we were not made miraculously on the sixth day. The earth is very old and we all came about through a process of evolution. That is true, and that is an advance in knowledge, whether or not you believe in a real world independent of observation. This is not to say that everything that happens in a scientific revolution is necessarily good or unambiguously better. In fact, later in the chapter I shall speak to these very issues and suggest that answers are complex. But there is no underlying problem here.

Having said this, even if you are dubious about scientific progress, I do not at all see why we should not look back with interest on the things that we see as affecting us today. And evolution certainly does. Nor do I see why we should not look back on things that we think true today. And I believe evolution certainly is. We should look back on these things and try to see how they relate to the present and how (if need be) we can now reverse or revise them. Admittedly, we should look back also on the things we think wrong, if only to see why we think them wrong and compare them to the things we think right. But one can do this in the case of the Darwinian Revolution. Indeed, many of us do precisely this. I have published a book on the history of American Creationism, comparing it with evolution and arguing that in many respects evolutionists and Creationists share the same problems and premises (Ruse 2005). It is legitimate for an evolutionist to look back on the history of vertebrates to try to understand why we humans have so many back problems today. It is no less legitimate for a historian to look back on the history of evolution to try to understand why we humans (in America especially) have so many science/religion conflicts today. Indeed, I would pitch the stronger claim that unless you are doing history with an eye to the present, you are simply indulging in a hobby and not scholarship. It is all a bit like building a model of St. Paul's Cathedral out of matchsticks. Fun but worthless.

So, all in all, the critics' worries are not well taken. Long live the evo-revo school!

Was there a Darwinian *Revolution?*

How much credit does Charles Robert Darwin merit for the revolution that carries his name? My writing this book and your reading it show that we both think he deserves some merit – a lot in fact. Before the *Origin of Species* appeared in 1859, the idea of evolution was a minority position and in many ways not very respectable. After the *Origin*, it became in many circles – middle-class and working-class, religious and not religious – the accepted position on origins. More than this – with a defiant nod at the critics of the last section, who do not want us to judge by the present – Darwin put forward the mechanism of natural selection, and today this is generally accepted as the right mechanism. Darwin got it right about causes.

But there is more to the question than this. Start with the period before Darwin. We now know, thanks to the mass of research by scholars, that there was greater acceptance of evolutionary ideas than we realized (Richards 2003). In Germany, that curious mélange of philosophers, writers, scientists, and even theologians, the *Naturphilosophen*, were generally not evolutionists in the sense of believing in actual links between successive forms. But it was never the links themselves, real or apparent, that mattered for them. It was much more the overall picture. Some certainly moved towards real evolutionism as the data came in. Even Goethe, toward the end of his long life, embraced the idea. In France, the opinion used to be that Lamarck was something of an oddity, and that Cuvier's anti-evolutionism was the universal norm. This was not strictly true. There was a whole group of evolutionists around Lamarck (Corsi 1988), and this continued through the century. We know of Etienne Geoffroy Saint-Hilaire, who upset Cuvier around 1830 with his evolutionism, but he was not alone (Laurent 1987). And there were others in other countries.

Britain, too, yields many evolutionists, starting with Charles Darwin's own grandfather, Erasmus. There has been a tendency rather to dismiss Erasmus as a fat fool, who was a bad poet and too much given to sexual pursuits. Now we realize that he had more influence than we knew. His major work, *Zoonomia*, was translated into German and read (and commented on) by the aged Immanuel Kant (Ruse 2006b). Charles Darwin himself had also read *Zoonomia*. Then, research in the past twenty years has shown how large a number of radical evolutionists there were in London in the 1830s. Particularly important was the professor of anatomy at University College London, Robert Grant, with whom (as an undergraduate in Edinburgh) Darwin had on many occasions discussed matters biological (Desmond 1989). There was even, for all of its detractors, *The Vestiges of the Natural History of Creation* (Secord 2000). The work was a major inspiration for the poet Alfred Tennyson as he struggled to finish (what rapidly became) his much-loved and -read poem *In Memoriam*. First, the poet read Lyell and worried that uniformitarian geology meant that everything is meaningless – it comes into being, stays a while, and then is swept away:

> Are God and Nature then at strife,
> That Nature lends such evil dreams?
> So careful of the type she seems,
> So careless of the single life; . . .

> So careful of the type? but no.
> From scarped cliff and quarried stone
> She cries, "A thousand types are gone:
> I care for nothing, all shall go."

Given nature "red in tooth and claw" – *In Memoriam* is the source of this famous phrase – nothing has meaning. Things just keep grinding along. But then Tennyson read Chambers, with his message of progress. All now made sense. Perhaps Arthur Hallam, a long-dead friend to whom the poem was dedicated, was a superior kind of being whose big mistake was to come before his time. In the cosmic scheme of things, all makes good sense.

A soul shall strike from out the vast
And strike his being into bounds,

And moved thro' life of lower phase,
Result in man, be born and think,
And act and love, a closer link
Betwixt us and the crowning race. . . .

Whereof the man, that with me trod
This planet, was a noble type
Appearing ere the times were ripe,
That friend of mine who lives in God.

All very odd if you take it too literally, but wildly inspirational for many of his fellow Victorians – including the most Victorian of them all, the queen on her throne – and evolutionary through and through.

We could go on listing pre-Darwinian evolutionists, most particularly the general man of letters and science Herbert Spencer, who in the 1850s was just beginning his dizzying rise upwards as the people's philosopher in Britain and the rest of the world. In the decade before Darwin, Spencer published evolutionary ideas, including a clear statement of natural selection (Ruse 1996; Richards 1987). So there can be no claim that Darwin was the first evolutionist or even the first to come up with natural selection. There were others who also had glimpses of selection, as well, of course, as Alfred Russel Wallace, whose sending to Darwin in 1858 of an essay that contained a clear expression of natural selection was the immediate spur to Darwin's writing the *Origin*.

Moreover, as I have stressed again and again, there is also the fact that Darwin rarely if ever had an original idea in his life. He was a great pack rat, forever gathering together the ideas of others. To get to natural selection, he had to learn all about artificial selection from the breeders – apart from some desultory experiments with pigeons, he was not involved in the practical aspects of any of this. He was not really involved in the theoretical aspects either, getting his information from others. Then there was the influence of Archdeacon William Paley (1802), who convinced Darwin that the world is design-like and that any natural mechanism for the creation of organisms had better take this fact into account. And let us not forget the crucial importance of Robert Malthus (1826), who argued that food supplies are outstripped by population pressures and that there will be ongoing struggles for existence. As we saw, this goes straight into the *Origin of Species*. "A struggle for existence inevitably follows from the high rate at which all organic beings tend to increase." Explicitly acknowledged: "It is the doctrine of Malthus applied with manifold force to the whole animal and vegetable kingdoms; for in this case there can be no artificial increase of food, and no prudential restraint from marriage" (Darwin 1859, 63).

More broadly – and again recapping material covered earlier in this book – there was the overall influence of Charles Lyell, whose *Principles of Geology* (1830–3) not only inspired Darwin in his early days as a geologist but whose general philosophy – of explaining the past by reference to causes now in operation – was the ruling method that Darwin used in his evolutionary theorizing. And complementing this was the influence first of the empiricist philosopher John F. W. Herschel (1830, 1841), whose writings probably drove Darwin to make so much in the *Origin* of the artificial–natural selection analogy. Plus the rationalist philosopher William Whewell (1837, 1840), whose argument that the best science shows a "consilience of inductions" – the bringing together of different areas of science under one causal hypothesis – is precisely the argument of the second half of the *Origin*. Darwin surveyed instinct, paleontology, biogeography, systematics, morphology, embryology, arguing that all of these are explained by evolution through selection, and in turn all of them make plausible evolution through selection. A classic consilience and Darwin was proud of the fact.

Romanticism?

One could keep going. There is today a ferocious debate about the influence on Darwin of German thought. Throughout this book I have emphasized the extent to which Darwin's thinking was played off against the ideas and visions of the *Naturphilosophen*, who saw life's processes as leading to a necessary progressive unfurling of ever greater complexity and sophistication. It would be disingenuous to conceal that there are those who take a completely opposite position. According to the eminent historian of science Robert J. Richards, Darwin was as close to a British *Naturphilosoph* as it was possible to be. Apart from anything else, supposedly a key factor in Darwin's intellectual development was the reading of the travel books of the German naturalist Alexander von Humboldt.

> Darwin came by his attitudes much as the earlier German Romantics had, through prolonged contact with exotic nature – but nature as filtered through a certain literature. In Darwin's case, the literature was singularly provided by the conceptually and aesthetically lush works of Alexander von Humboldt, who taught him how to experience the sublime and how morally to evaluate the nature he met in the jungles, mountains, and plains of South America. That early experience, formed and shaped under the guiding images provided by Humboldt, settled deeply into the conceptual structure of the *Origin of Species* and the *Descent of Man*. The sensitive reader of Darwin's works, a reader not already completely bent to early-twenty-first-century evolutionary constructions, will feel the difference between the nature that Darwin describes and the morally effete nature of modern theory.
>
> (Richards 2003, 552–3)

People like me counter this way of thinking about Darwin and his theorizing, arguing that it was Britain and its influences that really counted for the author of the *Origin*. In a letter to his friend Friedrich Engels, Karl Marx put the case exactly: "It is remarkable how Darwin recognizes among beasts and plants his English society with its division of labour, competition, opening up of new markets, 'inventions,' and the Malthusian 'struggle for existence'" (Letter, June 18, 1862; Marx 1981). I agree. I do not think you could have a more British scientific theory. If I have not convinced you of this by now, I never shall. But note that this is an argument about degree. No one denies that Darwin was influenced by German thinking, either directly or through others, notably Richard Owen and his theory of archetypes. It is more a question of whether Darwin started with the archetypes (Richards) or worked toward the archetypes (Ruse).

So, having poured so much water on the altar – having argued at length for the case that so many came before and that Darwin took and borrowed so much – what can one say in response to the question whether the revolution was a Darwinian one? How can one defend the Darwinian part of the Darwinian Revolution? Simply by pointing out that Darwin's genius was to take so many different ideas and to make something of them. Others set the problem. Evolution: right or wrong? The "mystery of mysteries." There may have been many evolutionists, but no one had come up with a way to make the idea plausible. It was very much in the realm of what one might call "pseudo-science," like phrenology and mesmerism. Darwin made the idea of evolution not just plausible but, for most folks, absolutely compelling. The way he tied everything together in a consilience was definitive – back then and now. Darwin did this in the *Origin*.

The same is true of natural selection. Darwin may have borrowed from others but it was he who made something of it all. Take Malthus. He used the struggle to argue that there can be no overall, lasting change. He argued that any attempts at state help – trying to improve the general status of humankind – were simply bound to make things worse. You feed the poor in this generation and you have more of them in the next. Darwin turned this on its head, showing how the struggle can make for ongoing change. Similarly with artificial selection, which was the standard argument against the possibility of ongoing change. You cannot turn horses into cows. In fact, in his essay, Wallace (1858) devoted much time to arguing that we should not take the analogy seriously and thus it is no bar to evolution. Again, Darwin saw the potential and used it – used it moreover to make an experienced case of change argue for unexperienced change, just the kind of argument that the empiricist Herschel was demanding.

Any fool can take pigments and paint a picture of flowers. It took Van Gogh to paint the sunflowers. It took Darwin to write the *Origin*. We can now move on to the time after the *Origin*. You might agree that Darwin put together the idea of evolution and made it compelling, and that it was because of him (together with his various supporters) that people were converted to it. However, drawing on earlier discussion, you may object that, after the *Origin*, natural selection was a flop. No one took it up, and it languished until the 1930s, when the population geneticists like Ronald A. Fisher (1930) in Britain and Sewall Wright (1931, 1932) in America melded Darwinian selection with Mendelian genetics to make the new theory, so-called "neo-Darwinism" or the "synthetic theory of evolution." Hence, in major respects, even if you grant that there was a revolution it was not very Darwinian. Indeed the historian Peter Bowler has gone so far as to write a book with the title, *The Non-Darwinian Revolution*! Adding to the negative case, we should also note that the way to any favorable reception of the *Origin* was mostly prepared by others. Theologians were important here. They were treating the Bible as if it were a human-written book – the approach known as "higher criticism" – and chipping away at literal interpretations of events described there. Literal readings of Genesis were discarded far less because of what Darwin said than because of what the church people themselves said.

As with the history of evolutionary theory before the *Origin*, there is today a major controversy about the history of evolutionary theory after the *Origin*. Peter Bowler (1996) thinks that good, professional-quality work was done – in Britain, Germany, France, and increasingly in America – and that this led to the synthetic theory. It was just that the work was not very Darwinian, in the sense of using natural selection. Others – and given the earlier discussions in this book, you will realize that this is my position – think that the years after the *Origin*, measured by the standards of mature professional science, were generally speaking an absolute disaster. It is not denied that there was some professional work, but it tended to the decidedly second-rate – phylogeny tracing – and it was increasingly out of touch with reality. The main use that was made of evolution by Darwin's supporters, like Thomas Henry Huxley, was as a Christianity substitute, a sort of secular religion, to promote the kind of society that they wanted to create. There was no wonder that Herbert Spencer, who was right at the forefront of this thinking, was more influential than Darwin.

Either way, it seems that the Darwinian Revolution was not very Darwinian. How does one respond? First by pointing out that it is not true to say that natural selection fell absolutely flat. After Darwin, Wallace (1870) and above all H. W. Bates (1862) did wonderful work on Lepidoptera and their markings, using selection as a tool of research. Then later in the century we find others also using selection – E. B. Poulton (1890) at Oxford and Raphael Weldon (1898) then in London, to name but two. Also, even if people did not wholly accept selection – and everyone thought it had some little role – the very fact of having a mechanism helped. It showed how things could be done, even if it was not itself a way of getting them done properly. But this kind of defence only goes so far. Really, one must agree that natural selection was not a great success. Scientists generally did not pick it up and use it. There was not a new field of selection studies. As we know, partly this non-development was scientific. There were perceived problems with selection – heredity and the age of the earth were issues that stood out. Partly, there were other factors for the non-development, some in science and some outside. Huxley, for example, was never really much interested in adaptation and design, so for him selection was not really needed anyway (Desmond 1994, 1997). Evolution was what he needed. Outside science, the religious were happy to accept evolution, but they still wanted a bit of divine guidance to get organisms, especially humans. So they were into directed mutations and so forth, and eschewed the full implications of the blind, cruel process of selection.

So, ultimately, let us agree that there was no Darwinian Revolution in this sense. But there is one fact that caps everything. Whatever people thought back then, Darwin did get it right about selection. It took seventy-five years for people to realize this, but then they did and now it is accepted. So, if you are prepared to use the present as a guide to the past – and I have defended this practice, so long as you do not gloss over how history took time to develop – I see no reason to deny Darwin's role in the Darwinian Revolution, and much good reason to think that the revolution is appropriately named. It was Charles Darwin who pointed to the causal approach taken by evolutionists today. Just as Mendel deserves credit for what he did, although he died unheralded by the scientific community, so Darwin deserves credit for what he did, although the scientific community of his day did not appreciate the strength of what he had done.

Was there a Darwinian Revolution?

We come to the final question and focus on the nature of the revolution: what kind of revolution was it really? In other words, the question is a philosopher's question about the nature of science and the way in which it changes. There are two basic positions here. We have

The Darwinian Revolution

had elements of them before, but let us recap. First there is the more conventional view, often associated with the name of Karl Popper (1959). This assumes that there is a real world and the aim of science is to get ever closer in understanding it – mapping it and explaining it. We may never get to our goal, we will certainly never be sure that we have reached our goal, but this is the aim nevertheless. It is a rational process of conjecture and test, and (where necessary) refutation. Old theories are either rejected for better ones or absorbed into newer ones (replacement or reduction). You can certainly speak of "revolutions," but these are not events outside the course of regular science. A revolution is simply more of what you have all of the time – bigger, more influential, but of the same type.

Then, second, there is the view over which with the name and work of Thomas Kuhn (1962) loom large. This says that the scientist works within a paradigm, a way of seeing things that in some sense determines reality itself. Most science is normal science within the paradigm, but every now and then there is a break, a shift, from one paradigm to another. It is not stupid to change paradigms – the old paradigm is no longer giving answers and throwing up new questions and challenges – but in some sense it is a-rational, outside reasoned argument. It is something of a different kind from the puzzle solving that goes on in normal science. It is much more like a political upheaval, and is rightly called "revolutionary." So let us phrase our question about the Darwinian Revolution in terms of these two views. Leave aside the ultimate metaphysical question about the reality of the world, which we discussed earlier and on which I am not sure that here we can make much headway, even assuming that there is headway to be made. Focus, rather, on the more epistemological issue about the nature of the revolution. Was the change smooth, as more conventional philosophies of science might lead one to expect? Was the change more Popperian, where new facts told against the older position and people shifted because this was the reasonable thing to do? Or was there a change of paradigms in the sense described by Thomas Kuhn in his *The Structure of Scientific Revolutions*? Was there a switch of world views – perhaps even a switch of worlds – that required more of a leap of faith than an appeal to reason?

After nearly forty years of looking at the revolution, my answer is unequivocal – yes and no (and maybe)! Taking a broad view, there are certainly Kuhnian aspects to the revolution. Most strikingly, there were people who simply could not see the other side's point of view – clever people, that is, who knew the ins and outs of the issues. Prominent among these was our old friend, the Swiss-American ichthyologist Louis Agassiz (Agassiz 1885). He had staked out an idealistic position before the *Origin* (Agassiz 1859) – one that came directly from his *Naturphilosoph* teachers (Friedrich Schelling and Lorenz Oken) when he was a student in Munich – and, try as he might, he could never accept evolution, even the Germanized form of it that was being pushed by people like Ernst Haeckel (1866). To Agassiz's credit, he really did try: his students around him, including his own son, were becoming evolutionists in the 1860s, but it was not for him.

This makes perfect sense in the Kuhnian scenario and fits uncomfortably into the Popperian (1959) scenario, although, in fairness to Popper, we should remember that he thought the Darwinian Revolution was more of a metaphysical shift than something purely scientific. He might well have thought in this case (as opposed, say, to the case of relativity theory) that it was not simply a matter of facts and reason, but more one of underlying commitments. So let us simply say, without trying to attribute positions to people, that someone like Louis Agassiz does not fit comfortably into a philosophy of science that makes rational choice the sole criterion of theory change.

However, in other respects the Darwinian Revolution seems clearly very non-Kuhnian. There have been major arguments about what precisely Kuhn meant by "different worlds." My own inclination is to say, on rereading the first edition of *The Structure of Scientific Revolutions*, that Kuhn meant a real ontological change, although tempered by the fact that ontology has to be mediated through the observer. He is more of a philosophical idealist than realist, thinking that the idea of a real world, independent of the observer, does not make a lot of sense. But, however you read Kuhn, he does argue that people see the facts differently: it is not just a question of interpretation, but of the facts themselves. This is simply not true of the Darwinian Revolution. Everything about Darwin himself denies this claim. Darwin was not like the Christian God, making things from nothing. He was much more like Plato's Demiurge, shaping what he already had. This applied to ideas as well as facts. Everyone knew about Malthus, but it was Darwin's genius to put the ideas into a theory of change rather than a theory that argued that change is impossible. Likewise, the facts of the successes of animal and plant breeders were well known. Again, it was Darwin's genius to make something of these facts. The same is true of so much else – the vaguely progressive fossil record, for instance, and the peculiarities of biogeography. Particularly important for Darwin were Ernst von Baer's discoveries in embryology. Darwin seized on the similarities of embryos and made this a key support for the arguments of the *Origin*.

Let me underline this point by drawing attention to the time after Darwin, and returning briefly to the students of Agassiz. It was notorious then – and it is a burden of historians of science now – that it was often absolutely impossible from reading their papers to tell if they had crossed the evolutionary divide or not. The best of them all was Alpheus Hyatt (1889), a first-class invertebrate paleontologist. Unfortunately brilliance does not always correlate with clarity. Hyatt was a foggy writer by anyone's standards – his papers drove Darwin to despair – and part of that fogginess is that one simply does not know where he stood on the issue of evolution. You can read through a dense paper on long-extinct marine organisms, and come out at the end not knowing if he thinks them related or not. They could all have been planted in the record by the Intelligent Designer. I do not know how else you can describe this phenomenon except by saying that the facts remained the same and the interpretation mattered. Note that this applies to the fact of evolution and not to the cause of evolution, especially not to natural selection. If we focus less on world changing and more on world interpretation, then we do have a rather Kuhnian situation (which, as I have pointed out, in the specific instance of the Darwinian Revolution, was probably also the position of Popper). If you take natural selection out of the picture – and most people at Darwin's time did – then the switch to evolution was in a way more of a metaphysical switch, a switch to a natural world, than simply one of science. Many went on doing exactly the same science as before. It was just that now they thought that evolution rather than God was the proximate cause. Ignoring selection and focusing just on the big picture, nothing much changed down at the coal-face. Imagine if you read a molecular biology paper of the late 1950s, and you could not tell if the author accepted the double helix! It would be impossible to do any molecular biology without being fully aware of the Watson and Crick breakthrough.

Natural selection

So it was not so much the case that the facts in the Darwinian Revolution were unimportant. People were very impressed by the information in the *Origin*. But there was more than just a rational-choice-powered switch. What about natural selection? The facts obviously counted highly here – the facts bound up with the values of science: the explanations that could be given, the predictions that could be made, the consistency with other theories, and so forth. One sees this again and again in the classic works – for instance, Fisher's *Genetical Theory of Natural Selection* (1930). He spent a lot of time showing how a Mendel-infused Darwinism can work, and then went on to discuss such issues as camouflage and mimicry, taking real-life examples. The same is true in America. Take Theodosius Dobzhansky. Although initially, in the first edition of his *Genetics and the Origin of Species* (1937), he did not give an overwhelming role to natural selection, by 1941 when he published the second edition he was moving to selection. Dobzhansky's switch came simply from the facts of variation that he was finding in his fruitflies, especially those in the wild. They showed cyclic, seasonal changes that he simply could not explain by other processes – genetic drift for instance – and so he moved to a more selectionist stance (Lewontin et al. 1981). A rational, fact-driven decision if ever there was one.

This is not to say that there were no extra-scientific factors involved. Ronald Fisher was a fanatically committed Englishman, and the heritage and glory of Charles Darwin was part and parcel of his commitment. (Fisher was a friend of, and fellow eugenicist with, Darwin's youngest surviving son, Major Leonard Darwin. Childless Leonard Darwin helped out the always underfunded, large Fisher family.) For Fisher, natural selection was more than just something justified by the facts. It also had deep cultural significance and, hence, was to be cherished. The Church of England, of which Fisher was always a deeply committed member, has ever had a weakness for the Pelagian heresy – that is, for preferring good works over faith. Fisher fit right in.

> There is indeed a strand of moral philosophy, which appeals to me as pure gain, which arises in comparing Natural Selection with the Lamarckian group of evolutionary theories. In both of these contrasting hypotheses living things themselves are the chief instruments of the Creative activity. On the

Lamarckian view, however, they work their effect by willing and striving only; but, on the Darwinian view, it is by doing or dying. It is not mere will, but its actual sequel in the real world, its success or failure, that is alone effective.

We come here to a close parallelism with Christian discussions on the merits of Faith and Works. Faith, in the form of right intentions and resolution, is assuredly necessary, but there has, I believe, never been lacking through the centuries the parallel, or complementary, conviction that the service of God requires of us also effective action. If men are to see our good works, it is of course necessary that they should be good, but also and emphatically that they should work, in making the world a better place.

(Fisher 1950, 19–20)

The moral to be drawn from all of this is that the reasons for scientific moves – especially the big ones – are tangled and not necessarily all of a kind. Both Popper and Kuhn had deep insights into the nature of scientific change, and although on the surface they seem diametrically opposed, in order to understand something like the Darwinian Revolution, you are better off if you draw from both sides rather than trying to fit the change wholly into one of the categories and excluding the other.

Form versus function

This reflection brings me to the final point to be made. Go back to the distinction made in an earlier chapter between form and function. Is the right way to understand organisms to regard them as having basic forms – archetypes or *Baupläne* (as the Germans called them) – with adaptations added on top? Or is the right way to understand organisms to regard them as being adaptive wonders (or machines) that show shared form because of shared ways in which adaptation works? No one denies either form or function as having a role, but which is prior? Is it final cause first and pattern second, or pattern first and final cause second? These two perspectives, form and function, suggest to me features of Kuhnian paradigms. The identification does not work completely, because all biologists recognize both simultaneously (Russell 1916). But form and function are paradigm-like in some respects, not the least being that they both involve metaphors, which Kuhn (especially in his later writings) saw as lying at the heart of paradigms. One can see organisms as being almost crystal-like – snowflake-like – with repetitions within and analogies without, and one can see organisms as being design-like, with the intricate and coordinating workings of the various parts. As central, or more central, is the question of commitment and vision. Some biologists see form as prior, believing that this is a basic commitment, and they simply cannot see why others do not share their vision. And some biologists see function as prior, believing that this is a basic commitment, and they simply cannot see why others do not share their vision.

In the *Origin*, Charles Darwin recognized both form and function – Conditions of Existence and Unity of Type – and clearly came down in favor of function over form. He thought that shared patterns were the consequence of evolution, and that the real driving force and issue was adaptation. It was to this that natural selection spoke. But now notice how these two paradigms/metaphors relate and how they persist in parallel rather than occurring sequentially (as Kuhn argued for paradigms). Leave Darwin's personal views to one side. The Darwinian Revolution was a failure if you think that historically it represents the victory of form over function. On this issue, there was no revolution. Before Darwin there were formalists. Goethe was one, the English biologist Richard Owen, with his theory of archetypes, was another. Before Darwin there were functionalists. Archdeacon Paley was one, the great French comparative anatomist Georges Cuvier was another. Most of these people – Paley and Cuvier for sure, and Goethe and Owen for much of their careers – were not evolutionists. If you need a formalist who was not an evolutionist, then add Louis Agassiz.

What is fascinating is that, after Darwin, there were (evolutionary) functionalists. Bates was one. Later there were Weldon and Fisher. Now there are Richard Dawkins (1986) – who describes himself as being somewhat to the right of Archdeacon Paley on the issue of adaptation – and many other evolutionists, including this author. But there were also (evolutionary) formalists. Darwin's bulldog, Thomas Henry Huxley, was one, and in Germany the promoter of *Darwinismus*, Ernst Haeckel, was another. After that we have, at the beginning of the twentieth century, people like the Scottish morphologist D'Arcy Wentworth Thompson (1917). And today we have Dawkins's great rival in the popular field, the late Stephen Jay Gould (2002). All of his arguments about spandrels were designed to promote form over function. Others of Gould's ilk include those who think that the laws of physics create form – the "order for free" school – including the Stuart Kauffman (1993) and the Canadian–Englishman Brian Goodwin (2001).

My gut instinct as a partisan is to say that one side is right and the other wrong. My more reflective opinion as a historian is that we have rival and ongoing world pictures, very much akin to Kuhnian paradigms. They are not strictly paradigms, for, apart from the fact that both sides see at least something in the opinions of the rivals, there is not a sequential process, with one paradigm ousting the other. Rather there are ongoing world pictures, and in this sense the Darwinian Revolution was no revolution.

Conclusion

Part of me feels distinctly queasy in drawing this book to an end on such a note. After everything has been said, the conclusion is that there was no Darwinian Revolution? Did I make a category mistake when I wrote a book on the topic? Of course that is not the conclusion of this book. I can make such an evaluation as I have just made mainly because I am supremely confident that what Darwin did was produce one of the most important works in human history. What I have been trying to do in this chapter is understand why the *Origin of Species* was one of the most important works in human history. It detracts not at all from Darwin and his achievements to put them into context and to discover what he did and did not do. So, let us end on a high note, with a clear statement of what Charles Robert Darwin did do. His theory of evolution through natural selection, as given in the *Origin of Species*, changes forever the way in which we look at the world and how we understand ourselves. This has implications in every area of human thought, including – especially including – philosophy.

Bibliography

Agassiz, E. C., ed. 1885. *Louis Agassiz: His Life and Correspondence.* Boston: Houghton Mifflin

Agassiz, L. 1859. *Essay on Classification.* London: Longman, Brown, Green, Longmans, and Roberts and Trubner

Allison, A. C. 1954a. Protection by the sickle-cell trait against subtertian malarial infection. *British Medical Journal* 1: 290

——. 1954b. The distribution of the sickle-cell trait in east Africa and elsewhere and its apparent relationship to the incidence of subtertian malaria. *Transactions of the Royal Society of Tropical Medical Hygiene* 48: 312

Atran, S. 2004. *In Gods we Trust: The Evolutionary Landscape of Religion.* New York: Oxford University Press

Augustine. 1998. *The City of God against the Pagans,* ed. and trans. R. W. Dyson. Cambridge: Cambridge University Press

Ayala, F. J. 1995. The myth of Eve: molecular biology and human origins. *Science* 270: 1930–6

Babbage, C. 1967. *The Ninth Bridgewater Treatise: A Fragment,* 2nd edn. London: Frank Cass [orig. pubd 1837]

Bakker, R. T. 1983. The deer flees, the wolf pursues: incongruencies in predator–prey coevolution, *Coevolution,* ed. D. J. Futuyma and M. Slatkin. Sunderland, Mass.: Sinauer

Bates, H. W. 1977. Contributions to an insect fauna of the Amazon valley, *Collected Papers of Charles Darwin,* ed. P. H. Barrett. Chicago: Chicago University Press, 87–92 [orig. pubd in *Transactions of the Linnean Society* 23, 1862]

Beecher, H. W. 1885. *Evolution and Religion.* New York: Fords, Howard, and Hulbert

Behe, M. 1996. *Darwin's Black Box: The Biochemical Challenge to Evolution.* New York: Free Press

Bernhardi, F. von. 1912. *Germany and the Next War*, trans. A. H. Powles. London: Edward Arnold

Bowler, P. J. 1976. *Fossils and Progress*. New York: Science History Publications

———. 1986. *Theories of Human Evolution*. Baltimore: Johns Hopkins University Press

———. 1988. *The Non-Darwinian Revolution: Reinterpreting a Historical Myth*. Baltimore: Johns Hopkins University Press

———. 1996. *Life's Splendid Drama*. Chicago: University of Chicago Press

Boyer, P. 2002. *Religion Explained: The Evolutionary Origins of Religious Thought*. New York: Basic Books

Brandon, R. M., and M. D. Rauscher. 1996. Testing adaptationism: a comment on Orzack and Sober. *American Naturalist* 148: 189–201

Brown, P., T. Sutikna, M. J. Morwood, R. P. Soejono, E. Jatmiko, E. Wayhu Saptomo, and Rokus Awe Due. 2004. A new small-bodied hominin from the Late Pleistocene of Flores, Indonesia. *Nature* 431: 1055–61

Browne, J. 1995. *Charles Darwin*, 1: *Voyaging*. New York: Knopf

———. 2002. *Charles Darwin*, 2: *The Power of Place*. New York: Knopf

Bumpus, H. C. 1898. The elimination of the unfit as illustrated by the introduced sparrow, *Passer domesticus*. *Biological Lectures from the Marine Biological Laboratory, Woods Hole, Mass.* 6: 209–26

Burchfield, J. D. 1974. Darwin and the dilemma of geological time. *Isis* 65: 300–21

Cain, A. J. 1954. *Animal Species and their Evolution*. London: Hutchinson

Cannon, W. F. 1961. The impact of uniformitarianism: two letters from John Herschel to Charles Lyell, 1836–1837. *Proceedings of the American Philosophical Society* 105: 301–14

Carroll, R. L. 1997. *Patterns and Processes of Vertebrate Evolution*. Cambridge: Cambridge University Press

Carroll, S. B. 2005. *Endless Forms Most Beautiful: The New Science of Evo Devo*. New York: Norton

Carroll, S. B., J. K. Grenier, and S. D. Weatherbee. 2001. *From DNA to Diversity: Molecular Genetics and the Evolution of Animal Design*. Oxford: Blackwell

Chambers, G. K. 1988. The *Drosophila* alcohol dehydrogenase gene-enzyme system. *Advances in Genetics* 25: 39–107

[Chambers, R.] 1844. *Vestiges of the Natural History of Creation*. London: Churchill

Committee on Strategies for the Management of Pesticide Pest Populations, 1986. *Pesticide Resistance: Strategies and Tactics for Management*. Washington, D.C.: National Academy Press

Conway-Morris, S. 2003. *Life's Solution: Inevitable Humans in a Lonely Universe*. Cambridge: Cambridge University Press

Corsi, P. 1988. *The Age of Lamarck*. Berkeley: University of California Press

Daly, M., and M. Wilson. 1988. *Homicide*. New York: De Gruyter

Darwin, C. 1859. *On the Origin of Species by Means of Natural Selection, or, The Preservation of Favoured Races in the Struggle for Life*. London: John Murray

——. 1862. *On the Various Contrivances by which British and Foreign Orchids are Fertilized by Insects, and On the Good Effects of Intercrossing*. London: John Murray

——. 1871. *The Descent of Man and Selection in Relation to Sex*. London: John Murray

——. 1887. *The Life and Letters of Charles Darwin, Including an Autobiographical Chapter*, ed. F. Darwin. London: John Murray

——. 1958. *The Autobiography of Charles Darwin, 1809–1882*, ed. Nora Barlow. London: Collins

——. 1959. *The Origin of Species by Charles Darwin: A Variorum Text*, ed. M. Peckham. Philadelphia: University of Pennsylvania Press

——. 1985–. *The Correspondence of Charles Darwin*, ed. F. Burkhardt and S. Smith. Cambridge: Cambridge University Press

——. 1987. *Charles Darwin's Notebooks, 1836–1844: Geology, Transmutation of Species, Metaphysical Enquiries*, ed. P. H. Barrett, P. J. Gautrey, S. Herbert, D. Kohn, and S. Smith. Ithaca, NY: Cornell University Press

Darwin, C., and A. R. Wallace. 1958. *Evolution by Natural Selection*, with foreword by Gavin de Beer. Cambridge: Cambridge University Press

Darwin, E. 1791. *The Botanic Garden: A Poem*. London: J. Johnson, pt 1: *The Economy of Vegetation*

——. 1794–6. *Zoonomia, or, The Laws of Organic Life*. London: J. Johnson [3rd edn. 1801]

——. 1803. *The Temple of Nature*. London: J. Johnson

Davies, N. B. 1990. Dunnocks: cooperation and conflict among males and females in a variable mating system, *Cooperative Breeding in Birds*, ed. P. B. Stacey and W. D. Koenig. Cambridge: Cambridge University Press, 455–85

Dawkins, R. 1976. *The Selfish Gene*. Oxford: Oxford University Press

——. 1982. *The Extended Phenotype: The Gene as the Unit of Selection*. Oxford: W. H. Freeman

——. 1983. Universal Darwinism, *Evolution from Molecules to Men*, ed. D. S. Bendall. Cambridge: Cambridge University Press

——. 1986. *The Blind Watchmaker*. New York: Norton

——. 1989. The evolution of evolvability, *Artificial Life*, ed. C. G. Langton. Redwood City, Calif.: Addison-Wesley, 201–20

——. 1995. *A River out of Eden*. New York: Basic Books

——. 1997. Human chauvinism: review of Stephen Jay Gould, *Full House*. *Evolution* 51, no. 3: 1015–20

——. 2003. *A Devil's Chaplain: Reflections on Hope, Lies, Science, and Love.* Boston and New York: Houghton Mifflin

——. 2006. *The God Delusion.* New York: Houghton Mifflin

Dawkins, R., and J. R. Krebs. 1979. Arms races between and within species. *Proceedings of the Royal Society of London* 205: series B, 489–511

Dembski, W. 2005. Intelligent design's contribution to the debate over evolution: a reply to Henry Morris. http://www.designinference.com/documents/2005.02.Reply_to_Henry_Morris.htm (accessed February 1, 2005)

Dennett, D. C. 1984. *Elbow Room: The Varieties of Free Will Worth Wanting.* Oxford: Oxford University Press

——. 1995. *Darwin's Dangerous Idea.* New York: Simon and Schuster

——. 2006. *Breaking the Spell: Religion as a Natural Phenomenon.* New York: Viking

Desmond, A. 1989. *The Politics of Evolution: Morphology, Medicine, and Reform in Radical London.* Chicago: University of Chicago Press

——. 1994. *Huxley, 1: The Devil's Disciple.* London: Michael Joseph

——. 1997. *Huxley, 2: Evolution's High Priest.* London: Michael Joseph

Dewey, J. 1909. *The Influence of Darwin on Philosophy.* New York: Henry Holt

Dick, S. J. 1996. *The Biological Universe: The Twentieth-Century Extraterrestrial Life Debate and the Limits of Science.* Cambridge: Cambridge University Press

Dobzhansky, Th. 1937. *Genetics and the Origin of Species.* New York: Columbia University Press

Dudley, J. W. 1977. Seventy-six generations of selection for oil and protein percentages in maize. *Proceedings of the International Conference on Quantitative Genetics,* ed. E. Pollak, O. Kempthorne, and T. B. Bailey. Ames, Ia.: Iowa State University Press, 459–73

Eddington, A. S. 1929. *Science and the Unseen World.* London: Macmillan

Eldredge, N., and S. J. Gould. 1972. Punctuated equilibria: an alternative to phyletic gradualism. *Models in Paleobiology,* ed. T. J. M. Schopf. San Francisco: Freeman, Cooper, 82–115

Elgin, M. 2006. There may be strict empirical laws in biology, after all. *Biology and Philosophy* 21: 119–34

Ellegård, A. 1958. *Darwin and the General Reader.* Göteborg: Göteborgs Universitet

Endler, John. 1986. *Natural Selection in the Wild.* Princeton: Princeton University Press

Engels, F. 1964. *The Dialectics of Nature, 1873–86,* 3rd edn., trans. C. Dutt. Moscow: Progress

Ereshefsky, M., ed. 1992. *The Units of Evolution: Essays on the Nature of Speciation.* Cambridge, Mass.: M.I.T. Press

Erskine, F. 1995. "The Origin of Species" and the science of female inferiority. *Charles Darwin's "The Origin of Species": New Interdisciplinary Essays.* Manchester: Manchester University Press, 95–121

Falk, D. 2004. *Braindance: New Discoveries about Human Origins and Brain Evolution.* Gainsville, Fl.: University of Florida Press

Farley, J. 1977. *The Spontaneous Generation Controversy from Descartes to Oparin.* Baltimore: Johns Hopkins University Press

Farlow, J. O., C. V. Thompson, and D. E. Rosner. 1976. Plates of the dinosaur Stegosaurus: forced convection heat loss fins? *Science* 192: 1123–5

Feduccia, A. 1996. *The Origin and Evolution of Birds.* New York: Yale University Press

Fisher, R. A. 1930. *The Genetical Theory of Natural Selection.* Oxford: Oxford University Press

——. 1950. *Creative Aspects of Natural Law: The Eddington Memorial Lecture.* Cambridge: Cambridge University Press

Fodor, J. 1996. Peacocking. *London Review of Books* no. 18, April: 19–20

Forrest, B., and P. R. Gross. 2004. *Creationism's Trojan Horse: The Wedge of Intelligent Design.* Oxford: Oxford University Press

Freud, S. 1964. The question of a *Weltanschauung. The Standard Edition of the Complete Psychological Works of Sigmund Freud,* trans. J. Strachey, 22: *New Introductory Lectures on Psycho-Analysis and Other Works.* London: Hogarth Press and Institute of Psycho-analysis, 158–84 [orig. pubd 1932–6]

Fry, I. 2000. *The Emergence of Life on Earth.* Newark, N.J.: Rutgers University Press

Gasman, D. 1971. *The Scientific Origins of National Socialism: Social Darwinism in Ernst Haeckel and the Monist League.* New York: Elsevier

Ghiselin, M. T. 1974a. *The Economy of Nature and the Evolution of Sex.* Berkeley: University of California Press

——. 1974b. A radical solution to the species problem. *Systematic Zoology* 23: 536–44

Giere, R. 1988. *Explaining Science: A Cognitive Approach.* Chicago: University of Chicago Press

Gilbert, S. F., J. M. Opitz, and R. A. Raff. 1996. Resynthesizing evolutionary and developmental biology. *Developmental Biology* 173: 357–72

Goodwin, B. 2001. *How the Leopard Changed its Spots,* 2nd edn. Princeton: Princeton University Press

Gould, S. J. 1980. Is a new and general theory of evolution emerging? *Paleobiology* 6: 119–30

——. 1988. On replacing the idea of progress with an operational notion of directionality, *Evolutionary Progress,* ed. M. H. Nitecki. Chicago: University of Chicago Press, 319–38

——. 1989. *Wonderful Life: The Burgess Shale and the Nature of History*. New York: Norton

——. 1996. *Full House: The Spread of Excellence from Plato to Darwin*. New York: Paragon

——. 2002. *The Structure of Evolutionary Theory*. Cambridge, Mass.: Harvard University Press

Gould, S. J., and R. C. Lewontin. 1979. The spandrels of San Marco and the Panglossian paradigm: a critique of the adaptationist program. *Proceedings of the Royal Society of London* 205: series B, 581–98

Grant, M. 1916. *The Passing of the Great Race, or, The Racial Basis of European History*. New York: Charles Scribner's Sons

Grant, P. R. 1986. *Ecology and Evolution of Darwin's Finches*. Princeton: Princeton University Press

Grant, R. B., and P. R. Grant. 1989. *Evolutionary Dynamics of a Natural Population: The Large Cactus Finch of the Galapagos*. Chicago: University of Chicago Press

Gray, A. 1881. *Structural Botany*, 6th edn. London: Macmillan

Haeckel, E. 1866. *Generelle Morphologie der Organismen*. Berlin: Georg Reimer

Haldane, J. B. S. 1949. *What is Life?* London: Lindsay Drummond

Hamilton, W. D. 1964a. The genetical evolution of social behaviour, I. *Journal of Theoretical Biology* 7: 1–16

——. 1964b. The genetical evolution of social behaviour, II. *Journal of Theoretical Biology* 7: 17–32

Hamilton, W. D., R. Axelrod, and R. Tanese. 1990. Sexual reproduction as an adaptation to resist parasites. *Proceedings of the National Academy of Science, USA* 87, no. 9: 3566–73

Harcourt, A. H., P. H. Harvey, S. G. Larson, and R. V. Short. 1981. Testis weight, body weight and breeding system in primates. *Nature* 293: 55–7

Hegel, G. W. F. 1970. *Philosophy of Nature*, trans. A. V. Miller. Oxford: Oxford University Press [orig. pubd 1817]

Hempel, C. G. 1966. *Philosophy of Natural Science*. Englewood Cliffs, N.J.: Prentice-Hall

Herbert, S. 2005. *Charles Darwin, Geologist*. Ithaca, N.Y.: Cornell University Press

Herschel, J. F. W. 1830. *A Preliminary Discourse on the Study of Natural Philosophy*. London: Longman, Rees, Orme, Brown, Green, and Longman

——. 1841. Review of William Whewell, *History of the Inductive Sciences* and *Philosophy of the Inductive Sciences*. *Quarterly Review* 135: 177–238

Hillis, D. M., J. P. Huelsenbeck, and C. W. Cunningham. 1994. Application and accuracy of molecular phylogenies. *Science* 264: 671–7

Hodge, M. J. S. 2005. Against "Revolution" and "Evolution". *Journal of the History of Biology* 38: 101–21

Holldöbler, B., and E. O. Wilson. 1990. *The Ants*. Cambridge, Mass.: Harvard University Press

Hrdy, S. B. 1981. *The Woman that Never Evolved*. Cambridge, Mass.: Harvard University Press

——. 1999. *Mother Nature: A History of Mothers, Infants, and Natural Selection*. New York: Pantheon Books

Hull, D. L., ed. 1973. *Darwin and his Critics*. Cambridge, Mass.: Harvard University Press

Hull, D. L. 1980. Individuality and selection. *Annual Review of Ecology and Systematics* 11: 311–32

——. 1988. *Science as a Process*. Chicago: University of Chicago Press

Hume, D. 1947. *Dialogues Concerning Natural Religion*, ed. N. K. Smith. Indianapolis, Ind.: Bobbs-Merrill [orig. pubd 1779]

——. 1963. The natural history of religion. *Hume on Religion*, ed. R. Wollheim. London: Fontana [orig. pubd 1757]

——. 1965. *A Treatise of Human Nature*, ed. L. A. Selby-Bigge. Oxford: Oxford University Press [orig. pubd 1739–40]

Huxley, J. S. 1912. *The Individual in the Animal Kingdom*. Cambridge: Cambridge University Press

——. 1942. *Evolution: The Modern Synthesis*. London: Allen and Unwin

——. 1943. *TVA: Adventure in Planning*. London: Scientific Book Club

Huxley, J. S., and J. B. S. Haldane. 1927. *Animal Biology*. Oxford: Oxford University Press

Huxley, T. H. 1858. On the theory of the vertebrate skull: Croonian Lecture delivered before the Royal Society, June 17, 1858, *Proceedings of the Royal Society* 9: 381–457

——. 1863. *Evidence as to Man's Place in Nature*. London: Williams and Norgate

——. 1877. *American Addresses, with a Lecture on the Study of Biology*. London: Macmillan

——. 1884. The Darwinian hypothesis, *Collected Essays: Darwiniana*. London: Macmillan, 1–21 [orig. pubd 1859]

——. 1989. *Evolution and Ethics with New Essays on its Victorian and Sociobiological Context*, ed. J. Paradis and G. C. Williams. Princeton: Princeton University Press

Hyatt, A. 1889. Genesis of the Arietidae. *Bulletin of the Museum of Comparative Zoology* 16, no. 3: 238

Irwin, D. E., S. Bensch, and T. D. Price. 2001. Speciation in a ring. *Nature* 409: 333–7

Isaac, G. 1983. Aspects of human evolution. *Evolution from Molecules to Men*, ed. D. S. Bendall. Cambridge: Cambridge University Press, 509–43

James, W. 1880. Great men, great thoughts, and the environment. *Atlantic Monthly* 46: 441–59

——. 1956. *The Will to Believe and Other Essays in Popular Philosophy and Human Immortality.* New York: Dover

——. 1967. What pragmatism means. *The Writings of William James: A Comprehensive Edition.* New York: Random House [orig. pubd 1904]

Jerison, H. 1973. *Evolution of the Brain and Intelligence.* New York: Academic Press

Johanson, D., and M. Edey. 1981. *Lucy: The Beginnings of Humankind.* New York: Simon and Schuster

Johnson, P. E. 1991. *Darwin on Trial.* Washington, D.C.: Regnery Gateway

——. 1995. *Reason in the Balance: The Case Against Naturalism in Science, Law, and Education.* Downers Grove, Ill: InterVarsity Press

Kant, I. 1928. *The Critique of Teleological Judgement,* trans. J. C. Meredith. Oxford: Oxford University Press [orig. pubd 1790]

Kauffman, S. A. 1993. *The Origins of Order: Self-Organization and Selection in Evolution.* Oxford: Oxford University Press

——. 1995. *At Home in the Universe: The Search for the Laws of Self-Organization and Complexity.* New York: Oxford University Press

Kelley, A. 1981. *The Descent of Darwin: The Popularization of Darwinism in Germany, 1860–1914.* Chapel Hill, N.C.: University of North Carolina Press

Kellogg, V. L. 1912. *Beyond War: A Chapter in the Natural History of Man.* New York: Henry Holt

Kettlewell, H. B. D. 1973. *The Evolution of Melanism.* Oxford: Clarendon Press

Kimura, M. 1983. *Neutral Theory of Molecular Evolution.* Cambridge: Cambridge University Press

King-Hele, D. 1963. *Erasmus Darwin: Grandfather of Charles Darwin.* New York: Scribners

Kitcher, P. 2003. *In Mendel's Mirror: Philosophical Reflections on Biology.* New York: Oxford University Press

Knoll, A. 2003. *Life on a Young Planet: The First Three Billion Years of Evolution on Earth.* Princeton: Princeton University Press

Kropotkin, P. 1955. *Mutual Aid.* Boston: Extending Horizons Books [orig. pubd 1902]

Kuhn, T. 1970. *The Structure of Scientific Revolutions,* 2nd edn. International Encyclopedia of Unified Science, 2/2. Chicago: University of Chicago Press [orig. pubd 1962]

Lakoff, G., and M. Johnson. 1980. *Metaphors We Live By.* Chicago: University of Chicago Press

Lamarck, J. B. 1809. *Philosophie zoologique.* Paris: Dentu

Larson, E. J. 1997. *Summer for the Gods: The Scopes Trial and America's Continuing Debate over Science and Religion.* New York: Basic Books

Laurent, G. 1987. *Paléontologie et évolution en France de 1800 à 1860: une histoire des idées de Cuvier et Lamarck à Darwin.* Paris: Éditions du C.T.H.S

Lennox, J. G. 2001. *Aristotle's Philosophy of Biology*. Cambridge: Cambridge University Press

Lewin, R. 1989. *Human Evolution: An Illustrated Introduction*, 2nd edn. Oxford: Blackwell Scientific

——. 1993. *The Origin of Modern Humans*. New York: Scientific American Library

Lewontin, R. C. 1974. *The Genetic Basis of Evolutionary Change*, Columbia Biological Series 25. New York: Columbia University Press

——. 1991. *Biology as Ideology: The Doctrine of DNA*. Toronto: Anansi

Lewontin, R. C., J. A. Moore, W. B. Provine, and B. Wallace, eds. 1981. *Dobzhansky's Genetics of Natural Populations I–XLIII*. New York: Columbia University Press

Locke, J. 1959. *An Essay Concerning Human Understanding*, ed. A. C. Fraser. New York: Dover [orig. pubd 1690]

Lorenz, K. 1982. Kant's doctrine of the "a priori" in the light of contemporary biology. *Learning, Development, and Culture: Essays in Evolutionary Epistemology*. Chichester: Wiley [orig. pubd 1941 as Kant's Lehre vom a priorischen im Lichte gegenwärtiger Biologie. *Blätter für Deutsche Philosophie* 15: 94–125]

Lumsden, C. J., and E. O. Wilson. 1981. *Genes, Mind, and Culture*. Cambridge, Mass.: Harvard University Press

Lyell, C. 1830–3. *Principles of Geology: Being an Attempt to Explain the Former Changes in the Earth's Surface, by Reference to Causes now in Operation*. London: John Murray

MacArthur, R. H., and E. O. Wilson. 1967. *The Theory of Island Biogeography*. Princeton: Princeton University Press

MacCulloch, D. 2004. *The Reformation: A History*. New York: Viking

McDonald, J. H., G. K. Chambers, J. David, and F. J. Ayala. 1977. Adaptive response due to changes in gene regulation: a study with *Drosophila*. *Proceedings of the National Academy of Sciences USA* 74: 4562–6

Mackie, J. 1977. *Ethics*. Harmondsworth: Penguin

McMullin, E. 1983. Values in science. *PSA 1982*, ed. P. D. Asquith and T. Nickles. East Lansing, Mich.: Philosophy of Science Association, 3–28

——. 1986. Introduction: evolution and creation. *Evolution and Creation*, ed. E. McMullin. Notre Dame, Ind.: University of Notre Dame Press, 1–58

McShea, D. W. 1991. Complexity and evolution: what everybody knows. *Biology and Philosophy* 6: 303–25

Majerus, M. E. N. 1998. *Melanism: Evolution in Action*. Oxford: Oxford University Press

Malthus, T. R. 1914. *An Essay on the Principle of Population*, 6th edn. London: Everyman [orig. pubd 1826]

Margulis, L. 1970. *Origin of Eukaryotic Cells*. New Haven, Conn.: Yale University Press

Bibliography

Marshall, L. G., S. D. Webb, J. J. Sepkoski Jr., and D. M. Raup. 1982. Mammalian evolution and the great American interchange. *Science* 215: 1351–7

Marx, K. 1981. *Karl Marx, Friedrich Engels: Selected Letters: The Personal Correspondence, 1844–1877*, ed. F. J. Raddatz, trans. E. Osers. Boston: Little, Brown

Maynard Smith, J. 1978. *The Evolution of Sex*. Cambridge: Cambridge University Press

——. 1981. Did Darwin get it right? *London Review of Books* 3, no. 11: 10–11

——. 1982. *Evolution and the Theory of Games*. Cambridge: Cambridge University Press

——. 1988. Evolutionary progress and levels of selection. *Evolutionary Progress*. Chicago: University of Chicago Press, 219–30

Maynard Smith, J., and E. Szathmáry. 1995. *The Major Transitions in Evolution*. New York: Oxford University Press

Mayr, E. 1942. *Systematics and The Origin of Species*, Columbia Biological Series 13. New York: Columbia University Press

——. 1954. Change of genetic environment and evolution. *Evolution as a Process*. London: Allen and Unwin, 157–80

——. 1969a. Commentary. *Journal of the History of Biology* 2: 123–8

——. 1969b. *Principles of Systematic Zoology*. New York: McGraw-Hill

Mayr, E., and W. B. Provine, eds. 1980. *The Evolutionary Synthesis: Perspectives on the Unification of Biology*. Cambridge, Mass.: Harvard University Press

Mill, J. S. 1874. *A System of Logic*, 8th edn. London: Longmans, Green

Miller, K. 1999. *Finding Darwin's God*. New York: Harper and Row

Miller, S. L. 1953. A production of amino acids under possible primitive Earth conditions. *Science* 117: 528–9

Mitchison, G. J. 1977. Phyllotaxis and the Fibonacci series. *Science* 196: 270–5

Mithen, S. 1996. *The Prehistory of the Mind*. London: Thames and Hudson

Moore, G. E. 1903. *Principia Ethica*. Cambridge: Cambridge University Press

Morris, H. M. 1989. *The Long War Against God: The History and Impact of the Creation/Evolution Conflict*. Grand Rapids, Mich.: Baker Book House

Morwood, M. J., and others. 2004. Archaeology and age of a new hominin from Flores in eastern Indonesia. *Nature* 431: 1087–91

Myers, F. W. H. 1881. "George Eliot", *Century Magazine*, November

Nagel, E. 1961. *The Structure of Science: Problems in the Logic of Scientific Explanation*. New York: Harcourt, Brace, and World

Newman, J. H. 1973. *The Letters and Diaries of John Henry Newman*, 25, ed. C. S. Dessain and T. Gornall. Oxford: Clarendon Press

Niklas, K. J. 1988. The role of phyllotactic pattern as a "developmental constraint" on the interception of light by leaf surfaces. *Evolution* 42: 1–16

Noll, M. 2002. *America's God: From Jonathan Edwards to Abraham Lincoln.* New York: Oxford University Press

Numbers, R. L. 1992. *The Creationists: The Evolution of Scientific Creationism.* New York: Knopf

——. 1998. *Darwinism Comes to America.* Cambridge, Mass.: Harvard University Press

Orzack, S. H., and E. Sober, eds. 2001. *Adaptationism and Optimality.* Cambridge: Cambridge University Press

Oster, G., and E. O. Wilson. 1978. *Caste and Ecology in the Social Insects.* Princeton: Princeton University Press

Owen, R. 1860. Review of Charles Darwin, *On the Origin of Species. Edinburgh Review* 111: 487–532

Paley, W. 1819a. *Collected Works*, 2: *The Principles of Moral and Political Philosophy.* London: Rivington [orig. pubd 1785]

——. 1819b. *Collected Works*, 3: *Evidences of Christianity.* London: Rivington [orig. pubd 1794]

——. 1819c. *Collected Works*, 4: *Natural Theology.* London: Rivington [orig. pubd 1802]

Pannenberg, W. 1993. *Towards a Theology of Nature.* Louisville: Westminster/ John Knox Press

Pasterniani, E. 1969. Selection for reproductive isolation between two populations of maize, *Zea mays L. Evolution* 23: 534–47

Peirce, C. S. 1877. The fixation of belief. *Popular Science Monthly* 12, November: 1–15

——. 1893. Evolutionary love. *The Monist*, 3: 176–200 [repr. in *Collected Papers of C. S. Peirce*, 6, ed. C. Hartshorne and P. Weiss. Cambridge, Mass.: Harvard University Press, 1935, 287–317]

——. 1958. The architecture of theories. *Values in a Universe of Chance: Selected Writings of Charles S. Peirce*, ed. P. P. Wiener. New York: Dover, 1958 [orig. pubd in *The Monist*, 1, 1891, 161–76]

Pennock, R. 1998. *Tower of Babel: Scientific Evidence and the New Creationism.* Cambridge, Mass.: M.I.T. Press

Plantinga, A. 1991. An evolutionary argument against naturalism. *Logos* 12: 27–49

——. 1993. *Warrant and Proper Function.* New York: Oxford University Press

——. 1998. When faith and reason clash: evolution and the Bible. *The Philosophy of Biology*, ed. D. Hull and M. Ruse. Oxford: Oxford University Press, 674–97 [orig. pubd in *Christian Scholar's Review* 21, no. 1: 8–32]

Polkinghorne, J. 1991. *Reason and Reality: The Relationship between Science and Theology.* Philadelphia: Trinity Press

Popper, K. R. 1959. *The Logic of Scientific Discovery.* London: Hutchinson

——. 1972. *Objective Knowledge.* Oxford: Oxford University Press

——. 1974. Darwinism as a metaphysical research programme. *The Philosophy of Karl Popper*, ed. P. A. Schilpp. La Salle, Ill.: Open Court, 1: 133–43

Poulton, E. B. 1890. *The Colours of Animals*. London: Kegan Paul, Trench, Trübner

Provine, W. B. 1971. *The Origins of Theoretical Population Genetics*. Chicago: University of Chicago Press

Putnam, H. 1981. *Reason, Truth and History*. Cambridge: Cambridge University Press

Quine, W. V. O. 1969. *Ontological Relativity and Other Essays*. New York: Columbia University Press

Rawls, J. 1971. *A Theory of Justice*. Cambridge, Mass.: Harvard University Press

Reeve, H. K., and L. Keller. 1999. Levels of selection: burying the units-of-selection debate and unearthing the crucial new issues. *Levels of Selection in Evolution*, ed. L. Keller. Princeton: Princeton University Press, 3–14

Reichenbach, B. R. 1976. Natural evils and natural laws: a theodicy for natural evil. *International Philosophical Quarterly* 16: 179–96

Reynolds, V., and R. Tanner. 1983. *The Biology of Religion*. London: Longman

Richards, R. J. 1987. *Darwin and the Emergence of Evolutionary Theories of Mind and Behavior*. Chicago: University of Chicago Press

——. 2003. *The Romantic Conception of Life: Science and Philosophy in the Age of Goethe*. Chicago: University of Chicago Press

Ridley, M. 1986. *Evolution and Classification: The Reformation of Cladism*. New York: Longman

Roberts, J. H. 1988. *Darwinism and the Divine in America: Protestant Intellectuals and Organic Evolution, 1859–1900*. Madison, Wisc.: University of Wisconsin Press

Rorty, R. 1995. Response to Bernstein. *Rorty and Pragmatism: The Philosopher Responds to his Critics*, ed. H. J. Saatkamp Jr. Nashville, Tenn.: Vanderbilt University Press

——. 2006. Born to be good: review of Marc Hauser, *Moral Minds*. *New York Times*, 27 August

Rudwick, M. J. S. 1969. The strategy of Lyell's *Principles of Geology*. *Isis* 61: 5–33

Ruse, M. 1973. *The Philosophy of Biology*. London: Hutchinson

——. 1975. Darwin's debt to philosophy: an examination of the influence of the philosophical ideas of John F. W. Herschel and William Whewell on the development of Charles Darwin's theory of evolution. *Studies in History and Philosophy of Science* 6: 159–81

——. 1979. *The Darwinian Revolution: Science Red in Tooth and Claw*. Chicago: University of Chicago Press [2nd edn. 1999]

——. 1980. Charles Darwin and group selection. *Annals of Science* 37: 615–30

———. 1986. *Taking Darwin Seriously: A Naturalistic Approach to Philosophy.* Oxford: Blackwell

———. 1987. Biological species: natural kinds, individuals, or what? *British Journal for the Philosophy of Science* 38: 225–42

———. 1993. Evolution and progress. *Trends in Ecology and Evolution* 8, no. 2: 55–9

———. 1996. *Monad to Man: The Concept of Progress in Evolutionary Biology.* Cambridge, Mass.: Harvard University Press

———. 1999. *Mystery of Mysteries: Is Evolution a Social Construction?* Cambridge, Mass.: Harvard University Press

———. 2000. *The Evolution Wars: A Guide to the Controversies.* Santa Barbara, Calif.: ABC-CLIO

———. 2001. *Can a Darwinian be a Christian? The Relationship Between Science and Religion.* Cambridge: Cambridge University Press

———. 2003. *Darwin and Design: Does Evolution Have a Purpose?* Cambridge, Mass.: Harvard University Press

———. 2005. *The Evolution–Creation Struggle.* Cambridge, Mass.: Harvard University Press

———. 2006a. *Darwinism and its Discontents.* Cambridge: Cambridge University Press

———. 2006b. Kant and evolution. *Theories of Generation*, ed. J. Smith, Cambridge: Cambridge University Press

Russell, B. 1945. *A History of Western Philosophy.* New York: Simon and Schuster

Russell, E. S. 1916. *Form and Function: A Contribution to the History of Animal Morphology.* London: John Murray

Russett, C. E. 1976. *Darwin in America: The Intellectual Response. 1865–1912.* San Francisco: Freeman

Sartre, J.-P. 1948. *Existentialism and Humanism*, trans. P. Mairet. London: Methuen [repr. 1977]

Secord, J. A. 2000. *Victorian Sensation: The Extraordinary Publication, Reception, and Secret Authorship of "Vestiges of the Natural History of Creation".* Chicago: University of Chicago Press

Sepkoski, D., and M. Ruse, eds. 2008. *The Paleobiological Revolution.* Chicago: University of Chicago Press

Sepkoski, J. J., Jr. 1978. A kinetic model of Phanerozoic taxonomic diversity, I: Analysis of marine orders. *Paleobiology* 4: 223–51

———. 1979. A kinetic model of Phanerozoic taxonomic diversity, II: Early Paleozoic families and multiple equilibria. *Paleobiology* 5: 222–52

———. 1984. A kinetic model of Phanerozoic taxonomic diversity, III: Post-Paleozoic families and mass extinctions. *Paleobiology* 10: 246–67

Sheppard, P. M. 1958. *Natural Selection and Heredity.* London: Hutchinson

Simpson, G. G. 1944. *Tempo and Mode in Evolution.* New York: Columbia University Press

——. 1962. *Principles of Animal Taxonomy.* New York: Columbia University Press

Skyrms, B. 1998. *Evolution of the Social Contract.* Cambridge: Cambridge University Press

——. 2002. Game theory, rationality, and evolution of the social contract. *Evolutionary Origins of Morality,* ed. L. D. Katz. Bowling Green, Ohio: Imprint Academic, 269–84

Sober, E. 1984. *The Nature of Selection.* Cambridge, Mass.: M.I.T. Press

——. 1988. *Reconstructing the Past: Parsimony, Evolution, and Inference.* Cambridge, Mass.: M.I.T. Press

Sober, E., and D. S. Wilson. 1997. *Unto Others: The Evolution of Altruism.* Cambridge, Mass.: Harvard University Press

Spencer, H. 1851. *Social Statics, or, The Conditions Essential to Human Happiness Specified and the First of them Developed.* London: J. Chapman

——. 1855. *Principles of Psychology.* London: Longman, Brown, Green, and Longmans

——. 1857. Progress: its law and cause. *Westminster Review* 67: 244–67

——. 1873. *The Study of Sociology.* London: Henry King

Stebbins, G. L., and F. J. Ayala. 1981. Is a new evolutionary synthesis necessary? *Science* 213: 967–71

Sterelny, K. 2007. The peculiar primate. *The Philosophy of Biology,* 2nd edn., ed. M Ruse. Buffalo, N.Y.: Prometheus

Sulloway, F. 1979. *Freud: Biologist of the Mind.* New York: Basic Books

Szostak, J. W., D. P. Bartel, and P. L. Luisi. 2001. Synthesizing life. *Nature* 409: 387–90

Thompson, D. W. 1917. *On Growth and Form.* Cambridge: Cambridge University Press

——. 1948. *On Growth and Form,* 2nd edn. Cambridge: Cambridge University Press

Tooby, J., L. Cosmides, and H. C. Barrett. 2005. Resolving the debate on innate ideas: learnability constraints and the evolved interpenetration of motivational and conceptual functions. *The Innate Mind: Structure and Content,* ed. P. Carruthers, S. Laurence, and S. Stich. New York: Oxford University Press, 305–37

Toulmin, S. 1967. The evolutionary development of science. *American Scientist* 57: 456–71

Tutt, J. W. 1891. *Melanism and Melanochroism in British Lepidoptera.* London: Swan Sonnenschein

Vermeij, G. J. 1987. *Evolution and Escalation: An Ecological History of Life.* Princeton: Princeton University Press

Wachtershauser, G. 1992. Groundwork for an evolutionary biochemistry: the non-sulfur world. *Progress in Biophysics and Molecular Biology* 58: 85–201

Wallace, A. R. 1858. On the tendency of varieties to depart indefinitely from the original type. *Journal of the Proceedings of the Linnean Society, Zoology* 3: 53–62

——. 1870. *Contributions to the Theory of Natural Selection: A Series of Essays.* London: Macmillan

——. 1876. *The Geographical Distribution of Animals.* London: Macmillan

——. 1900. *Studies: Scientific and Social.* London: Macmillan

——. 1905. *My Life: A Record of Events and Opinions.* London: Chapman and Hall

Watson, J. D., and F. H. C. Crick. 1953. Molecular structure of nucleic acids. *Nature* 171, 737–8

Wavell, S., and W. Iredale. 2004. Sorry, says atheist-in-chief, I do believe in God after all. *Sunday Times*, 12 December, section 1: 7

Weldon, W. F. R. 1898. Presidential address to the Zoological Section of the British Association. *Transactions of the British Association* 887–902

Whewell, W. 1833. *Astronomy and General Physics*, Bridgewater Treatise 3. London: William Pickering

——. 1837. *The History of the Inductive Sciences.* London: Parker

——. 1840. *The Philosophy of the Inductive Sciences.* London: Parker

Wilford, J. N. 2006. Fossil called missing link from sea to land animals. *New York Times*, 6 April, section A: 1, col. 5

Williams, G. C. 1966. *Adaptation and Natural Selection.* Princeton: Princeton University Press

Wilson, D. S. 2002. *Darwin's Cathedral.* Chicago: University of Chicago Press

Wilson, E. O. 1975. *Sociobiology: The New Synthesis.* Cambridge, Mass.: Harvard University Press

——. 1978. *On Human Nature.* Cambridge, Mass.: Harvard University Press

——. 1980. Caste and division of labor in leaf-cutter ants (Hymenoptera: Formicidae: *Atta*), 1: The overall pattern in *Atta sexdens*. *Behavioral Ecology and Sociobiology* 7: 143–56

——. 1984. *Biophilia.* Cambridge, Mass.: Harvard University Press

——. 1992. *The Diversity of Life.* Cambridge, Mass.: Harvard University Press

——. 2002. *The Future of Life.* New York: Vintage Books

——. 2006. *The Creation: A Meeting of Science and Religion.* New York: Norton

Wong, K. 2003a. An ancestor to call our own. *Scientific American* 13, no. 2: 4–13

——. 2003b. Who were the Neanderthals? *Scientific American* 13, no. 2: 28–37

Wright, S. 1931. Evolution in Mendelian populations. *Genetics* 16, no. 2: 97–159 [repr. in *Evolution: Selected Papers*, ed. W. B. Provine. Chicago: Chicago University Press]

——. 1932. The roles of mutation, inbreeding, crossbreeding and selection in evolution. *Proceedings of the VI International Congress of Genetrics*, 1932, 1: 356–66 [repr. in *Evolution: Selected Papers*, ed. W. B. Provine. Chicago: Chicago University Press]

Index

Page numbers in *italic* refer to illustrations.

adaptation
  complexity 150–4
  design metaphor and 72
  morphology 142–6
  optimality models 145–6
  rapid change 149–50
African Queen (film) 239
Agassiz, Louis 35
  cladism 140
  embryology 43, 45, 51
  formalist not evolutionist 306
  fossilization 41
  never accepted evolution 302,
    303
  pragmatists and 193
  threefold parallel 110
agriculture *see* artificial selection
Animal Biology (J. Huxley and
    Haldane) 180
ants, Wilson and 146
apes
  human evolution and 166–7,
    167
  relation to humans 159–60
  testicle size and 143–4
Aquinas, St Thomas 256

Aristotle 71, 289
  "entelechy" 101
artificial selection 21–4, 77
  hybrids 39–41
  variation of young 52
astronomy 125
Atran, Scott 278–9
Augustine of Hippo, St 246, 250,
    256
Autobiography (Darwin) 242
Ayala, Francisco 80, 126

Babbage, Charles 57, 243
Baer, Karl Ernst von 51, 303
barnacles 48
Barrett, H. C. 207
Bates, Henry Walter 15–16, 75–6,
    299
HMS *Beagle* voyage 242
  distribution patterns 7–9
  geological studies and 4, 5–7
  natives taken to England 18
  publication of 12–13
Beecher, Reverend Henry Ward
    245–6
bees 26, 38

Behe, Michael: *Darwin's Black Box*
258–9
Bentham, George 66
Bentham, Jeremy 216
Bergson, Henri 101
biogeography
American exchange 130–2
ecology and evolution 122–7
MacArthur–Wilson island theory
131–2
in *Origin* 45–7
plate tectonics and 127–32, *128,
129*
"ring of races" 123–5, *124*
birds
Archaeopteryx 115, *115*
artificial selection of *22,* 22–3
biogeography 46
classification of 136–7
Davies and dunnocks 96
Galapagos populations 8, 67–8,
79
rapid change 149–50
"ring of races" 123–5
sexual selection *28*
The Blind Watchmaker (Dawkins)
181–3, 188
Boulton, Matthew 1
Bowler, Peter: *The Non-Darwinian
Revolution* 299
Boyer, Pascal 277–8
*Breaking the Spell: Religion as a
Natural Phenomenon* (Dennett)
276–7
Bryan, William Jenning 249
Bumpus, Herman C. 76–7
Burke, Edmund 218
Burton, Richard 224
butterflies and moths
mimicry 75–6
studies of English moths 78–9

Calvin, John 247, 250, 275
Cambridge University 3, 4, 9–10
Candide (Voltaire) 148
capitalism: social Darwinism 221–2
Carnegie, Andrew 221
Catholic Church
acceptance of evolution 245
Hume on transubstantiation 267
miracles and 149, 255–6
causation
final-cause thinking 125
Greek philosophers and 71
metaphor of design 71–3
natural selection and 64–8
vera causa 64–8, 75
Chambers, G. K. 80
Chambers, Robert: *Vestiges of the
Natural History of Creation*
12, 100
Christianity
creation and evolution 245–9
Darwin's background 241–3
effect of Darwin 264
historical conflicts of 281–2
Huxley's view of 17
literal reading of Bible 246–9
as a meme 280–1
miracles 149, 242, 255–7
morality and 237
cladism 197–8
classification/systematics
cladism 137–42, *139,* 197–8
defining "species" 133–5
epistemological approaches
202–3
evolutionary theory and 132–7
Linnaean hierarchy *133*
metaphors for 69–70
molecular comparisons 141–2
in *Origin* 47–9
recent approaches 197–8

competition: struggle for existence
24–6; *see also* progress
complexity, genetics and 188–9;
see also design
convergence 183
Conway-Morris, Simon 183–4,
186, 187
Cosmides, L. 207
creationism 245–9
culture and knowledge 208
Crick, Francis 82, 94, 111
Critique of Teleological Judgement
(Kant) 72–3
culture
knowledge and 207–9
religion as 280–2
Cuvier, Georges 73, 294, 306

Darwin, Charles Robert
Beagle voyage 4–9
image as an ape *161*
inclined toward individuals 104
on language 196
on morality 215–19, 229–31,
232, 261
personal life
childhood and education 1–3
death of 20
friendships 12
ill-health of 11–12
marriage and children 11–12,
206
religious beliefs of 18, 241–3,
252, 268–71
reputation and respect 1,
293–300, 307
scientific methodology 56–8
works of
see also The Descent of Man;
*Origin of Species by Means
of Natural Selection*
Autobiography 242

The Descent of Man 158–60
joint papers with Wallace
12–13
*The Structure and Distribution
of Coral Reefs* 5–6, *6*
see also Darwinian revolution;
evolution; natural selection
Darwin, Dr Robert (father) 1, 2
Darwin, Emma Wedgwood 11–12
Darwin, Erasmus (brother) 3, 243
Darwin, Erasmus (grandfather) 1–2
on origin of life 100
progress of life 175–6
Zoonomia 8, 294
Darwin, Major Leonard (son) 304
Darwinian revolution
credit to Darwin for 293–300
effects of 300–3
form versus function 305–7
revolutionary aspects of 287–93
The Darwinian Revolution (Ruse)
287–90
"Darwinism and Philosophy"
(Dewey) 192
Darwin's Black Box (Behe) 258–9
Darwin's Dangerous Idea (Dennett)
286
Davies, Nicholas 96
Dawkins, Richard
on atheism 254
The Blind Watchmaker 181–3,
188
cleverness of selection 259
complex adaptation 150, 187,
244
concept of progress 181–3
epistemology and 210
on the evolutionary mechanism
263
The Extended Phenotype 186–7
functionalism 306
on human knowledge 209

memes 276
  on religion 279, 280, 283, 284,
    286
  "selfish gene" 109
  on suffering in nature 261
Dembski, William 260
Dennett, Daniel 235–6, 266
  Breaking the Spell 276–7
  Darwin's Dangerous Idea 286
  God as incoherent idea 283–4
  religion as culture 280–2
  religion as parasite 279
The Descent of Man (Darwin)
  evolutionary epistemology 196
  idea of progress 178–9
  influence of Hume 268–71
  man as animal 158–60
  morality and 215–19, 229–31
  sexual selection 18–20, 160–5
  works of others and 160
design
  argument from 253–5
  Intelligent Design 257–60, 265
  metaphors of 71–3
  problem of evil 260–3
Dewey, John: "Darwinism and
    Philosophy" 192
*Dialogues Concerning Natural
    Religion* (Hume) 253, 264
dinosaurs
  genetic cause and 116
  Stegosaurus form/function
    144–5, 148–9
disease and illness 77, 90–3
Disraeli, Benjamin 13
*Dissertation on the Progress of
    Ethical Philosophy* (Mackintosh)
    218
distribution of species *see*
    biogeography; populations
divergence 31, 33
diversity, geological *118*

Dobzhansky, Theodosius 96
  Genetics and the Origin of Species
    304
dogs 23, 51–2
Doyle, Arthur Conan
  Sign of Four 253–4
  Silver Blaze 99
Driesch, Hans 101
Dubois, Eugene 165

ecology, Haeckel's concept of
    25–6
Eddington, Arthur 61–2
Edwards, Milne 31
Eldredge, Niles 119–21
Eliot, George 268
elk, Irish 149, *150*
embryology 67
  Agassiz and 43
  concept of progress and 182
  homeotic genes and development
    154–7
  in *Origin* 51–2
  paleontology and 55, 110–11
Engels, Friedrich 297
environment, human intervention in
    79
*An Essay Concerning Human
    Understanding* (Locke)
    205–6
Essay on a Principle of Population
    (Malthus) 10–11
eukaryotes 112–15
*Evidence as to Man's Place in
    Nature* (T. Huxley) 160
Evidences of Christianity (Paley)
    241–2
evil, problem of
  free will versus determinism
    262–3
  Intelligent Design and 260–3
  moral versus physical 261–2

evolution
    as an ideology 16–17
    Beagle voyage and 7–9
    concept of progress 175–7
    as design 253–5
    Erasmus Darwin and 2
    hypo-deductive system 56–7
    initial controversies 13–16
    Malthus and 10–11
    mechanisms of 73–4
    molecular biology and 95–8
    Newton and 9–10
    other thinkers on 12
    as religion 299
    scientific method of 124–6
    "synthetic theory of" 20
    term of 13
    see also natural selection
Evolution: The Modern Synthesis
    (J. Huxley) 180
Evolutionarily Stable Strategy (ESS)
    107
The Extended Phenotype (Dawkins)
    186–7
eyes, vertebrate 34–5

Fibonacci sequence 152
fish
    paleontology 115
    phases of embryos 35
Fisher, Ronald A. 96–7, 298, 304,
    306
Flew, Anthony 101
Franklin, Benjamin 169
Frazer, J. G.: *The Golden Bough* 272
free will versus determinism 262–3
Fresnel, Augustin 66
Freud, Sigmund 272
fruitflies (*Drosophila*) 79–80, 116
    molecular embryology 154–6,
    156
The Future of Life (Wilson) 225–6

Galapagos Islands
    biogeography of 8–9, *9, 46*–7
    birds of 58–9, 136–7
    "ring of races" 123–4
Galileo Galilei 56
Galton, Samuel 1
Gärtner, Karl Friedrich von 40
genetics
    changing understanding of 20
    complexity and DNA 188–9
    Dawkins's "selfish gene" 109
    discovery of DNA 82, 94
    dominance 85
    "genetic drift" 97
    genotypes/phenotypes 83
    heredity and "pangenesis"
    14–15
    Mendelian 81–90
    molecular 94–8, 141–2
    mutations 85
    pleiotropy 211–12
    populations 86–90
    reductionism 94–5
    RNA 103
    shared structures 154–6, *156*
    sickle-cell anemia 90–3
    see also Mendel, Gregor; species
geographical distribution *see*
    biogeography; populations
geology
    "catastrophists" 4–6
    Darwin's focus on the *Beagle*
    4–7
    plate tectonics 127–32, *128*
    "uniformitarians" 5
    see also Lyell, Charles
Ghiselin, M. T. 109
Goethe, Johann Wolfgang von 35,
    293–4, 306
The Golden Bough (Frazer) 272
Goodwin, Brian 153, 307
Gould, John 136–7

Gould, Stephen Jay 119–21,
    151
  functionalism 306–7
  progress and 185–6, 251–2
  religion as a byproduct 277
  "spandrels" 146–8, 209
A Grammar of Assent (Newman)
    255
Grant, Peter and Rosemary 79
Grant, Robert 294
Gray, Asa 243, 261

Haeckel, Ernst 16, 302
  biogenetic law 52
  concept of "ecology" 25–6
  functionalism 306
  "Tree of Life" 32
Haldane, J. B. S. 102
  Animal Biology (with Huxley)
    180
Hallam, Arthur 295
Hamilton, William 105–6, 108
Hardy, G. H.: Hardy–Weinberg law
    87–90
Hegel, G. W. F. 35, 177
Hennig, Willi 138
Henslow, John 3, 12
heredity *see* genetics; natural
    selection
Herschel, John F. W. 8, 54, 296,
    298
  classification 136
  empiricism 64–6
  Preliminary Discourse 57
Hinton, Martin 166
Hitler, Adolf: *Mein Kampf* 224
Hodge, Jonathan 288
Hooker, Joseph 12–13
horses, paleontology of 115–16
Housman, A. E. 261
Hrdy, Sarah Blaffer 174
Hull, David 197–9, 200, 202–3

human beings
  Darwin sees as animal 158–60
  factors of evolution 169–70
  feeding big brains 170–1, 186,
    209
  fossil evidence of 165–9
  hobbits/*Homo floresiensis* 168–9,
    170, 186
  infanticide 163, 173–5
  intervention in nature 79
  language and 168
  male–female differences 162–4,
    172–4
  Neanderthals and 165
  progress of 175–84
  races of 164
  sociobiology and 171–5
  technology and 181–3
Humboldt, Alexander von 297
Hume, David 203
  *Dialogues Concerning Natural
    Religion* 253, 264
  on free will 262
  morality and 218, 219, 226,
    232, 237, 238
  The Natural History of Religion
    264, 265–8
  "Of the reason of animals" 219
  on religion 278, 283
Huxley, Julian
  Animal Biology (with Haldane)
    180
  Evolution: The Modern Synthesis
    180
  evolutionary ethics 225, 227
  *The Individual in the Animal
    Kingdom* 179–80
Huxley, Thomas Henry 12, 61, 74,
    228, 271
  approach to epistemology 195
  breeding and selection 67
  classification and 49

Huxley, Thomas Henry (*cont'd*)
  compared to Darwin 291
  creating species 80
  doubts progress 185
  education in science 17
  *Evidence as to Man's Place in
    Nature* 160
  evolution as religion 299
  functionalism 306
  metaphysical shift 111
  Neanderthals and humans 165
  paleontology 43, 115–16
  "saltationism" 15
Huygens, Christiaan 66
Hyatt, Alpheus 303
hybridity 39–41
Hymenoptera 105–6, *106*
hypo-deductive (H-D) system
    56–7
  natural selection argument
    58–61

idealism 193
*The Individual in the Animal
    Kingdom* (J. Huxley)
    179–80
instincts
  flexibility and 235–6
  individuals and groups 104–7
  of insects 235
  "optimality model" 36–9
  sympathy 219
intelligence
  feeding brains 170–1, 186, 209
  flexibility over instinct 235–6
  only one of many traits 189
  technology and 181–3
  see also knowledge
Intelligent Design theory 257–60
  Hume and 265
  problem of evil 260–3
Islam 276–7

James, William 173
  pragmatism 191–2
  response to Darwin 193–5
Jerison, Harry 181
Johnson, Philip 257–8
Journal of the History of Biology
    289
Just So Stories (Kipling) 148, *148*

Kant, Immanuel 10, 204
  Critique of Teleological Judgement
    72–3
  Darwin and 268
  final-cause thinking 125
  Metaphysics of Morals 216, 217
  pragmatists and 193
Kauffman, Stuart 152, 307
Kellogg, Vernon 223
Kettlewell, H. D. B. 78
Kimura, Motoo 98
Kipling, Rudyard: *Just So Stories*
    148, *148*
Knoll, A. 114
knowledge
  critique of pragmatism
    198–203
  culture and biology 207–9
  inductive reasoning 204–5
  innate capacities 203–7
  Kuhn's scientific paradigms
    201–3
  Plantinga and "Darwin's Doubt"
    210–14
  Platonic constructions 209–10
  pragmatism 191–203
Kropotkin, Prince Peter 222–3,
    228
Kuhn, Thomas S. 209
  form and function 305–6
  scientific paradigms 201–3
  *The Structure of Scientific
    Revolutions* 301–3, 305

Lamarck, Jean-Baptiste and
    Lamarckism 8
alternative to natural selection
    15, 16
complex adaptation 150
Darwin doesn't deny 26
evolutionary thought 294
first life 100–1
Freud on religion and 272
language
    concept of progress 177
    human evolution and 168
Leakey family 166
Leibniz, Gottfried W. 263
Lennox, J. G. 289
Lepidoptera *see* butterflies and
    moths
Lewontin, Richard 96, 262
life
    as an action 102
    as a force 101–2
    Miller–Urey experiments 102–3
    origins of 100–3
    RNA molecules 103
Linnaean Society of London 13
Linnaeus, Carolus 49, *133*
Locke, John: *An Essay Concerning
    Human Understanding* 205–6
Lorenz, Konrad 204
Lumsden, Charles 205
Lunar Society 1
Luther, Martin 247
Lyell, Charles
    ancient oceans 42
    arranges Wallace–Darwin
        publication 12–13
    Deism 243
    distribution of organisms 7, 8
    Herschel's praise for 65
    influence on Darwin 12, 56,
        296
    origins of humans 18

Principles of Geology 5–7, 242,
    296

MacArthur, Robert 122, 131–2
Mackie, John 240
Mackintosh, James: *Dissertation on
    the Progress of Ethical Philosophy*
    218
McShea, Dan 187–8
Majerus, M. E. N. 78
malaria and sickle-cell anemia 90–3
Malthus, Thomas Robert 298
    Essay on a Principle of Population
        10–11
    influence on Darwin 58, 244,
        296, 302
    Marx on 297
    struggle for existence 25
mammals
    continental drift and 129, 130
    saber-toothed 183, *184*
Margulis, Lynn 112–13
Marx, Karl and Marxism
    on Darwin and Malthus 297
    layered complexities 109
    life studies 102
Mayr, Ernst 121
    on reductionism 94
    species concepts 134–5
    systematics 197
    *Systematics and the Origin of
        Species* 123, 135
Mein Kampf (Hitler) 224
Mendel, Gregor 20, 73–4, 92,
    206, 300
    concept of progress and 179
    Darwinian context 81–3
    dominant genes 85
    law of transmission 83–4, *84*, 86
    the locus of alleles 84–5
    molecular genetics and 94–5
    reliability of data 85

metaphors 68–73
Metaphysics of Morals (Kant) 216,
    217
Mill, John Stuart 65
    reaction to *Origin* 54
    The Subjection of Women 164
Miller, Stanley 102
mimicry 77
    in Lepidoptera 75–6
molecular biology *see* genetics
Moore, G. E. 226
morality
    Darwin on 215–19, 229–31,
        232
    desires 217
    eugenics 225
    evolutionarily stable strategy
        (ESS) 234–5
    Hume on 267–8
    importance of 236–7
    justifying evolution ethics 226–8
    metaethics 215, 238–40
    naturalistic fallacy 226
    non-human behavior and 235–6
    normative 215, 216–19, 238–40
    political conflict 223–4
    preserving biodiversity 225–6
    recent theories 231–6
    social contract theory 232–4
    social Darwinism 219–23
    status of women 224–5
    sympathy 219
Morgan, Thomas Hunt 82, 85
morphology
    classification 136–7
    convergence 183
    form and function debate 142–6
    organized complexity 150–4
    rapid change 149–50
    "spandrels" 146–8, *147*
    spirals 152, *153*
    Unity of Type 49–50, *50*

moths *see* butterflies and moths
Muller, H. J. 85, 96
Müller, Max 196
Murray, John 21
mutation 85

The Natural History of Religion
    (Hume) 264, 265–8
natural selection 55
    alternatives to 15–16
    artificial selection and 21–4,
        77–80
    bringing on uniformity 30
    classification and 47–9
    Darwin's metaphors 68–9
    Deistic God and 244–5
    epistemological approach
        191–201
    individuals and groups 104–10
    isolation and 30–1
    James on 193–4
    jumps in fossil records 119–21
    mechanisms of 26–30
    neo-Darwinist evidence 77
    population genetics 86–90
    r and K selection theory 274–5
    reaction to idea 14–15
    scientific basis of 61–4
    struggle for existence 24–6
    as true cause/*vera causa* 64–8
    using hypo-deductive system
        58–61
    see also Darwinian revolution;
        genetics; species; *The Descent
        of Man*; *Origin of Species by
        Means of Natural Selection*
Natural Theology (Paley) 241,
    243
naturalism
    Intelligent Design theory
        257–60
    theology of 255

Naturphilosophen
  classification 49
  Darwin's use of 157
  embryology 51
  influence on Darwin 293–4,
    296–7
  origin of life and 100
  paleontology and 43
  patterns 35–6
  perfection in nature 177–8, 252
Newman, John Henry 246
  A Grammar of Assent 255
Newton, Isaac 20
  development of theory 9–10
  H-D system 56
  reality of gravity 136
  true causes 64–5
  Whewell on 57
Niklas, K. J. 153
The Non-Darwinian Revolution
  (Bowler) 299

Oken, Lorenz 35, 302
On Human Nature (Wilson)
  272–4
One Flew Over the Cuckoo's Nest
  (film) 238
Oparin, A. I. 102
*Origin of Species by Means of
    Natural Selection* (Darwin)
  287
  evidence for 75
  final paragraph of 52–3
  first formation of life 100–1
  idea of progress 178
  immediate reception of 13–14
  on mechanism of evolution 293
  Mendel and 82
  originality of 298–9
  publication and reception of
    13–16, 21, 54–5
  structure of argument 55–8

style of writing 55, 68–71
  writing of 11–13
  see also biogeography; natural
    selection; populations; species
Owen, Richard 49, 297
  on Darwin's writing style 68
  metaphysics 111
  theory of archetypes 306

paleontology
  after Darwin 110–18
  age of the earth *113*
  attitudes toward cause 116–17
  contemporary fossil record *44*
  diversity over time *118*
  embryology and 51, 55, 110–11
  human links 165–9
  jumps in fossil records 119–21
  linking species 115–16
  mass extinctions 119
  in *Origin* 41–5
  process of fossilization 41–2
  single-celled organisms 112–15
Paley, William 253, 254, 296, 306
  Evidences of Christianity 241–2
  Natural Theology 241, 243
  *The Principles of Moral and
    Political Philosophy* 241
Pasteur, Louis 100, 112
Peirce, Charles Sanders 62–3,
    192–3, 199
pesticide resistance *77*
phyllotaxis 152, *153*
Pictet, Marc-Auguste 41, 43
Piltdown hoax 166
Plantinga, Alvin 101, 255
  "Darwin's Doubt" 210–14
plants
  continental drift and 129, 130
  phyllotaxis 152, *153*
plate tectonics, biogeography and
    127–32

Plato 71
  cave of *The Republic* 213
  ideas from the soul 204
pleiotropy 211–12
Polkinghorne, Revd Sir John 283
Popper, Karl 63–4, 126
  evolutionary epistemology 201
  mystery of life 103
  rational process of science
    301–3, 305
  simplicity and 140–1
  theory of falsifiability 196
populations
  biogeography 121–7
  "genetic drift" 97
  genetics of 86–90
  individuals versus groups 104–10
  isolation and 30–1
  sex ratios 107–8
  sickle-cell anemia 90–3
Poulton, E. B. 299
pragmatism
  critique and responses to
    198–203
  response to Darwinism 191–8
*Preliminary Discourse on the Study
    of Natural Philosophy*
    (Herschel) 57
Priestley, Joseph 1
Principles of Geology (Lyell) 5, 242,
    296
*The Principles of Moral and
    Political Philosophy* (Paley) 241
Principles of Psychology (Spencer)
    203
progress
  concept of 175–7, 187–90
  Darwinists and 179–84
  Darwin's idea of 177–8
  doubts concerning 185–7
  perfection in nature 177–8
  Spencer's predetermination 194

prokaryotes 112–15
Protestant churches 246–9
Pusey, Edward 246

Quakers 221, 247
Quine, W. V. O. 204–5

Rawls, John 237
  social contract theory 232–4
reductionism 94–5
Reichenbach, B. R. 263
religion 263–4
  atheism 242, 254
  biology of belief 282–6
  as a byproduct 277–9
  conflicting beliefs 285–6
  creation and evolution 245–9
  Darwin's beliefs 241–5, 252
  Darwin's Humean discussion
    268–71
  Deistic God 242–5
  design argument 72–3, 253–5
  evolution as 299
  group-selective analysis 275–6
  historical perspective on 281–2
  human nature and 171–2
  Hume on natural religion 264,
    265–8
  Intelligent Design argument
    257–60
  interventions in nature 111
  miracles 149, 255–6
  morality and 237, 239
  naturalistic account 284–5
  as a parasite 276–7
  progression and 250–2
  reaction to *Origin* 13–14
  social Darwinism 221
  social science studies of 271–5
  the soul 271
reptiles, continental drift and 129,
    130

The Republic (Plato) 213
Reynolds, Vernon 274, 280
Richards, Robert J. 221, 297
Rorty, Richard 192, 238
Ruse, Michael 297
  The Darwinian Revolution
    287–90
Russell, Bertrand 145–6, 195

saltation 37
  complex adaptation 151
  Peirce and 193
Sartre, Jean-Paul 237
Schelling, Friedrich 35, 302
science
  education in 16–17
  empirical reality of 60–1
  hypo-deductive system 56–7, 95
  Kuhn's paradigms 201–3, 301–3
  logic of evolution 124–6
  metaphysical shift 111
  overthrow of Mendel 95
  Popper and rational process
    301–3
  statistical analysis in 62–3
  true causes 64–8
  value-free 189–90
  Victorian context 297–300
  see also Darwinian revolution
Scopes, John Thomas 248, 249
Sedgwick, Adam 3, 5, 12, 41
selection see natural selection
Sepkoski Jr, J. John 117, 123, 189
sexual selection
  attraction and 117
  The Descent of Man 160–5
  individuals versus groups 108
  intra-specific struggle 104
  introduced in Origin 27–8
  sex and 172–4
  testicle size and 143–4
sickle-cell anemia 125

Sign of Four (Doyle) 253–4
Silver Blaze (Doyle) 99
Simpson, G. G. 117, 134, 197
Skyrms, Brian 234–5
Smith, Adam 218
Smith, John Maynard 108
snails 97
Sober, Elliott 323
social behavior
  advantages of 169
  human beings 162–3, 171–5
  individuals versus groups 104–10
  reciprocal altruism 229–31
  religion and 273–5
  selfishness and altruism 109–10
  social contract theory 232–4
Social Darwinism 17, 220–1
sociobiology 104, 272–4
Sociobiology: The New Synthesis
  (Wilson) 104
species
  complexity of 187–8
  creating separate 80–1
  defining 133–4
  distribution of 45–7
  evolutionary systematics 132–7
  gel electrophoresis 96
  individual differences 37
  molecular genetics and 96
  progression 251–2
  sex and 137
  shared structures 154–6, 156
  struggle for existence 24–6
  variation 33–6, 55
Spencer, Herbert 194, 199, 295,
  299
  concept of progress 176–7
  Principles of Psychology 203
  social Darwinism 220–1
  "survival of the fittest" 27
spontaneous generation theory 100
Sterelny, Kim 171–2

Strabo 266
The Structure of Scientific Revolutions (Kuhn) 301–3, 305
The Subjection of Women (Mill) 164
survival of the fittest 27, 63–4, 68–9
systematics *see* classification
Systematics and the Origin of Species (Mayr) 123, 135

Tanner, Ralph 274, 280
technology 181–3
Tennyson, Alfred, Lord: *In Memoriam* 294–5
Thatcher, Margaret 220, 227
Thompson, D'Arcy Wentworth 151, 306
tigers 188, *188*
titanotheres 117, *117*
Tooby, J. 207
Toplady, Augustus Montague: "Rock of Ages" 247–8
tortoises 8–9, *9*
"ring of races" 123–4
Toulmin, Stephen 196–7
"Tree of Life" metaphor *32*, 69–70
Tutt, J. W. 78–9

Uganda 91
Unitarianism 221, 243, 269
United States, literal Bible reading in 248–9
Urey, Harold 102
utilitarianism 215–16

value-free science 189–90
values *see* morality
Vestiges of the Natural History of Creation (Chambers) 12, 100
Voltaire: *Candide* 148

Wallace, Alfred Russel
breeding and selection 67
feminism of 224–5
groups and selection 104–5
joint publication with Darwin 12–13
morality and evolution 228
natural selection 298, 299
sexual selection and 19
socialism and biology 222
spur to Darwin 295
"survival of the fittest" 27
Wallace's Line 127, *127*
Watson, James 82, 94, 111
Wedgwood, Emma *see* Darwin, Emma Wedgwood
Wedgwood, Josiah 2
Wedgwood family, religion and 243, 244–5
Weinberg, Steven 280
Weinberg, Wilhelm: Hardy-Weinberg law 87–90
Weldon, Raphael 299, 306
Wells, H. G. 16–17
whales 187–8, *188*
Whewell, William 12, 296
at Cambridge 3
classification 136, 189
"consilience of inductions" 64–5
geological controversies 5
hostile to *Origins* 54
influence on pragmatists 193
methodology 57–8
Wilson, David Sloan 232, 275–6, 277, 280, 281
Wilson, Edward O. 110
concentration on ants 146
concept of progress 180–1
ecology and evolution 122–3
epigenetic rules of thought 205
evolutionary ethics 226–7

The Future of Life 225–6
human behavior 172–3
On Human Nature 272–4
island biogeography 131–2
naturalistic religion 284–5
on religion 272–4, 282
Sociobiology: The New Synthesis
    104
Withering, William 1
Wittgenstein, Ludwig xi–xii

wolves 29–30
women
    male–female differences 162–4,
        172–4
    status of 224–5
Wright, Chauncey 152
Wright, Sewall 298
    genetic drift 96–7, *98*

Zoonomia (E. Darwin) 8, 294